A record spinner's memoirs...

The true story of the day-to-day, play-by-play adventures of a young, insecure, teenage wallflower, from a small Midwestern factory town, who dives head first in to the rabbit hole of psychedelia, flaxen, waxen, sex, drugs, and rock and roll.

In the dawning of the Age of Aquarius, Sean Conrad experiences the ups and downs of having just too damn much fun—and the consequences that come with it. From hanging out with rock stars to being homeless...and back again. From Porsches to the potholes of life. From scribbling autographs to signing divorce papers. Can there possibly be a happy ending?

KSFX was in the ABC-FM group, as was my Chicago station, WDAI-FM. Group president Allen Shaw had always liked me and had respect for my work ethic. He had tried to get me to stay at WDAI but understood I couldn't turn down an offer like KHJ. I phoned him now and explained how I had made a huge mistake going to work for Paul Drew. I went into the details of KPOI and told him I was back on the mainland and wanted to apply for the KSFX job.

"If it were totally up to me, Sean, I'd hire you back in a minute," he said. "Let me call the new general manager, Don Platt, and tell him about you and your situation. Check back with me in a few days."

Before I had a chance to get back to Allen, I got a call from Don Platt.

"Allen Shaw thinks I should interview you for the PD opening," he said. "Can you catch a flight to San Francisco in the next day or so? We'll pay for the flight up."

Thank God for that. My alternative would have been hitch-hiking, and I probably would have done it.

I got the job. And hummed, "San Francisco, Here I Come," all the way back to LA.

After twelve years of an on-again-off-again marriage, Beth and I finally admitted it was over. We agreed it would be foolish for us to go to San Francisco together just to break up again. She had made new friends in LA, and we decided it was time to get a divorce. We had gotten married too young, and now, tough as it was, we both looked forward to a new life apart.

I felt terrible leaving my daughters behind but assured them as soon as I got an apartment in the city, I'd fly them up regularly. I left the car with Beth and started sending her child support right away. With a heavy heart, I said goodbye to my girls, and with nothing but a suitcase, moved everything I owned to the "City by the Bay."

My San Francisco E-Ticket Ride was about to begin.

KUDOS for *Kickin' Out the Jams*

A page-turner that reads like a novel. Sean Conrad has written an entertaining and inspiring book for anyone who wants to experience those early days of rock radio. His story of sex, drugs, rock 'n' roll, and transformation is as uplifting as it is outrageous. *– Bonnie Hearn Hill, best-selling author*

A fascinating and humorous read! It is a well written book, and a sometimes zany look at one man's journey during the golden age of radio and music. From playing the hits to getting the hits played, Sean's adventures in the radio and record business are a wild ride! I could not put this down! I LOVED IT. Joel Newman, Rock & Roll Hall of Fame, Cleveland, Ohio http://catalog.rockhall.com/catalog/ARC -0276

In the days of Boss Radio, Sean Conrad had it all, lost it all, and got it back again, several times. KICKIN' OUT THE JAMS is funny, sad, ironic, and entertaining—sometimes on the same page. For anyone wondering what life was like behind the mic, you won't be able to put it down. *– John Ostlund, Owner, KYNO*

This is a great book. It reads like a conversation with friends at a bar. Probably the best book on radio that I have read since SuperJock by Larry Lujack. *– Jason Remington, SeaTacMedia.com, Seattle, Washington*

Sean Conrad's KICKING OUT THE JAMS is the 50,000 whats and 50,000 what-nots of radio. A wild nostalgic journey of destruction and disc jockeys. Sean takes you back to those golden days when radio was our constant companion. If you think what came out it was fun, wait till you see what went into it. This book is my pick hit of the year. *– Ken Levine, writer for TV shows MASH, Cheers, Frasier, The Simpsons, and named one of the best 25 blogs of 2011 by* Time Magazine.

Sex and drugs and rock n' roll aren't just a saying, it was a living for Sean Conrad. Dive into this book, and immerse yourself in radio and records, when the industry was truly fun, funny, and the stuff great stories are made of. – *Laurie Roberts San Francisco Bay Area Radio Hall of Famer*

If Jack Kerouac's *On The Road* had been about the golden age of radio and rock and roll, it would be this book. – *Larry April, Senior Partner, April/ Vaughn Advertising, Santa Cruz, CA*

If you're a radio person, Sean Conrad's book *Kickin' Out the Jams* will make you remember the fun, crazy, irreverent, mostly magical times we had. If you weren't privileged to work in a radio station, this book will take you to places and events you'd never otherwise experience. Sean's stories epitomize the oft-stated radio adage: "You can tell the quality of the jock by the size of the U-Haul he brings to town." – *BC Cloutier, Former radio General Manager*

Wow! From cover to cover, *Kickin' Out The Jams* tells the story of one of the greatest eras in radio. Whether you're in the broadcasting industry or not, I guarantee you'll be glued to every page. Read Sean's book and you'll see why "The hits just kept on comin'!" You better believe It, baby! – *"Shotgun Tom" Kelly, Afternoon Radio Personality, KEARTH Radio, Los Angeles*

KICKIN' OUT THE JAMS

THE PURPLE HAZE OF MY CRAZY DAZE IN RADIO

Sean Conrad

A BLACK OPAL BOOKS PUBLICATION

Black Opal Books
BECAUSE SOME STORIES JUST HAVE TO BE TOLD

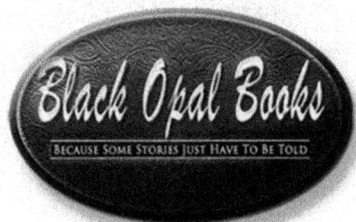

GENRE: NON-FICTION/MEMOIRS/HUMOR

KICKIN' OUT THE JAMS: The Purple Haze of My Crazy Daze in Radio
Copyright © 2013 by Ron Copeland
All Rights Reserved
Cover Design by Jackson Cover Designs and Spotfarm Studios
© 2013 All Rights Reserved
Print ISBN: 978-1-626940-06-2

First publication: APRIL 2013

Published by Black Opal Books: **http://www.blackopalbooks.com**

DEDICATION

*To my wife, Lisa, who had to live and re-live the
good and the bad parts of my life,
over and over and over again. Without her support,
this book would only be a dream.*

PROLOGUE

Goldie Hawn was the first person I saw when I walked into Lou Adler's mansion that night. Coming out of a small bathroom just off the foyer, she was dressed in only a huge, oversize football jersey. The black marks under her eyes made her look like a parody of a pro football player.

In the main living room, I found myself shoulder-to-shoulder with bigger than life stars like Warren Beatty, Sonny, Cher, and Jim Brown. My wife, Beth, and I stood next to a table full of exotic food, holding drinks and facing each other. Just behind Beth, John Lennon conversed with his current girlfriend, Mai Pang, who interestingly enough resembled Yoko Ono.

We were two of the very few people at the party without a star on the Hollywood Walk of Fame. This was truly an honest to god private, inside the movie industry holiday gathering so darting from room-to-room with an autograph book was not an option.

"Beth, we've just got to try and meet him," I told my wife. "If we don't, we'll regret it for years to come."

"How do we do it without being obvious?" she asked.

We stood there trying to get up the courage to pounce, knowing he could wander away to another room any second.

"Okay, let's just brush by him on our way to the next room, and I'll say something like, '*Oops, sorry.*'"

The room was wall-to-wall people. A few steps, a hesitant stumble, and the deed was done.

"John, my name is Sean Conrad and this is my wife, Beth. I'm the program director of KHJ Radio." Everyone knew the power of KHJ, including John Lennon, who had a solo career going and needed air play like any other artist.

"Nice to meet you," he replied in proper British accent.

I'd done it. It was a once-in-a-lifetime event for a kid from Beavercreek, Ohio. I remember thinking, *Wow, I've really made it.* How I got there is another story…

CHAPTER 1

"LONELY BOY"

Before the CD, before the iPod, before audio streaming, music was reproduced on a record player. A 45 was a 7-inch record with a big hole in the middle. It rotated at 45 rounds per minute. A 78 was a 10-inch record with a small hole in its middle that rotated at 78 rounds per minute. A 33 1/3 was a 12-inch record with a small hole in the middle that rotated at 33 1/3 rotations per minute. I was born in 45 and turned 33 in 78. I guess you could say I was born to play records.

I feel lucky to be alive. I almost died a few times from a disease called Too Much Fun, with a heavy dose of self-destruction thrown in for good measure. In my early childhood, I lived in several homes that had no indoor plumbing. We had to use a pump to draw water and then heat it over a flame, in order to take a hot bath. In the middle of a freezing night, I would grab a flashlight, throw on some warm winter clothes, and stumble through the snow to relieve myself in a drafty outhouse. The toilet was a hole in the ground with a splintery wooden seat over it. In one house, a coal-burning stove in the middle of the tiny living room was the only thing that kept us warm. We were poor, but I don't remember it being a bad thing.

We moved frequently. From the first grade through my sophomore year, I never attended the same school for a complete year. I went to two or three different schools each one of those years. Every time we moved, I once again had to be walked in to another new room filled with kids, most of whom had known each other since their diaper days. With the smell of polished wooden floors, mimeograph fluid, text books, and "today's special" cooking down the hallway, yet another new teacher would march me in front of the room already in progress and say, "Now class, I want you to meet our new student. His name is Ronnie Copeland. Please make him comfortable here."

"Hey, kid, you're ugly."

"Look, he's got holes in his shoes!"

Ducking spitballs, I was led to a desk like Ralphie had in *Christmas Story*, the kind that had the kid in front of you sitting on a bench seat that was connected to the front of your desk. You cannot imagine the utter fear I felt while trying to ignore the cat-calls and giggles. I'd have to go through the agony of proving myself to another new group of kids. That included the mandatory getting my ass kicked a few times by one of the school bullies. After a few bloody noses, I was eventually accepted into the fold only to get up and move again. And the whole gut-wrenching process would start all over. Feeling like a nobody during those years gave me one thing, and that was a strong desire to succeed.

Finally, late in my sophomore year, we moved to Beaver-creek Ohio, a sleepy little suburb of Dayton that became the ideal setting for me to get anchored and start working on my low self-esteem. During the summer between my sophomore and junior years, I remember sitting in my room alone, thinking that by the time I graduated from Beaver Creek High School, everyone would know who I was, by God. That is, if we didn't move again. And, we didn't.

I made the football team that year, was elected to the student council, became a home room representative, and won parts in the junior and senior class plays. My date for the senior prom Pam Robinson went on to become Miss Ohio in 1967, and I took the home coming queen Janet Zimmerman to the home coming dance. In short, I became an over-achiever. The final step in my quest for recognition was to get into radio.

When I was fourteen years old, I loved rock and roll. Instead of doing homework, I'd listen to Little Richard, Jerry Lee Lewis, Elvis Presley, and Chuck Berry on WING. At night, I'd listen on my rocket radio. It was a small plastic rocket-to-the-moon-shaped radio with a silver rod on its nose that you could pull in and out to tune in an AM radio station. It didn't need electricity. I just hooked a wire that came out of the bottom to the metal bed springs. It worked as an antenna, pulling in radio signals that I could listen to with a tiny earphone. My parents had no idea that instead of sleeping, I was secretly under the covers listening to Bob Holliday on WING and periodically calling in a request. I was always blown away when he'd actually play the song I asked for.

Before I was old enough to drive, one Saturday afternoon I begged my mom to drive me to downtown Dayton in her black and white, two-door hardtop, 1955 Chevy so I could buy a record. In December of 1960, the number one song on the charts was "Shop Around" by Smokey Robinson and the Miracles, which was their first number one hit. Every week, Mayors Jewelers at the corner of 3rd and Main streets featured the "record of the week" at a discount. That particular week it was "Shop Around" on Motown Records and I loved it. For forty-nine cents and six Pepsi bottle caps, you could own it. My mom's side of the family guzzled Pepsi, so bottle caps I had.

As I was climbing in the back seat of the car to go home, I accidentally broke half of the record. But that didn't matter. If you placed it on the turntable just right, it played just fine. I still have it somewhere in my collection of old 45s.

When I was in the eighth grade, my dad's younger brother, Jack, opened a small, used furniture store and soon grew it in to the largest store in Dayton. Jack was a shrewd businessman and knew how to sell furniture. To encourage foot traffic, he arranged for Bob Holliday to broadcast live on Sunday afternoons from the showroom window of the store. Bob was a 26- year-old radio announcer from San Angelo, Texas. He was always impeccably dressed in a custom-tailored, sharkskin suit with a pencil-thin neck tie held in place with a gold tie tack. He sported a neatly manicured crew-cut and drove a brand new white Cadillac convertible, loaded with options.

Bob was always chewing on a fingernail and could get very moody at times. He was a loner, and I don't think he was all that happy. He did have a strong work ethic and, in that department, was a great role model. I learned the basics of radio from him.

"Hi, this is Bob Holliday, and we're broadcasting live from Time Furniture and Appliances, located right next to the Varsity Bowling Alley on South Main Street here in Dayton. Stop on by, and the first five people who do will get a free copy of the latest record by local favorites, Teddy & The Rough Riders. And while you're here, take advantage of our broadcast special. Three full rooms of furniture for only $275! But hurry, this offer ends today at 6 p.m."

My father was a sales rep for my uncle. He called me at home the first day Bob was to appear and said, "Son, why don't you come down to the store and meet Bob Holliday?"

"Bob Holliday? I listen to him all the time."

"I'll bet he'll give you a free record," Dad said.

"I'm on my way."

I disliked going to the store because of my part-time job there. My uncle would take in used refrigerators and stoves and re-sell them. Most of the time, they came in filthy. Grease all over the stove top, mold in the refrigerators. Uncle Jack would pay me three dollars for each one of these smelly, dirty appliances that I cleaned up for re-sale. I hated it, but I liked being independent and having my own money.

And now, I got to meet Bob Holliday. Me? I couldn't believe it. I hurried down to the cross-town city bus, that ran on electric cables suspended from telephone poles throughout the city, and went to the store.

The bus dropped me off across the street but right in front of the showroom window of the furniture store. From there, I could see Bob talking into a microphone that was positioned between two turntables. Without waiting for the *Walk* signal to flash, I darted across the street and into the store. My dad stood at the door waiting for me.

"Hi, son. You made it. Let me take you to meet Bob. He only has another half-hour of his show left."

As we came around a row of new recliners toward Bob's set up, I was filled with anticipation about meeting my favorite DJ.

"Mr. Holliday, this is my son, Ronnie Copeland. He listens to you all the time."

"Uh hi, Mr. Holliday, I uh listen to you all the time," I repeated awkwardly.

"Well hello, Ronnie. Would you like a copy of this week's Top 30 Music Guide?"

"Thanks. Uh could you play "Mack The Knife" for me? I really like Bobby Darin."

Bob played "Mack The Knife," and that was the beginning. He took a liking to me, and any time he was broadcasting from the furniture store, I would be sitting close by, helping out in any way I could. He became my mentor into radio, and I never looked back. From the first day I hung out at Time Furniture and Appliances, watching Bob talk into that microphone, I never had a doubt in my mind as to what I wanted to do with my life.

WING in downtown Dayton, Ohio in the 1960s

Bob Holliday Publicity Shot

A WING survey featuring
Lou Swanson

CHAPTER 2

"DREAM LOVER"

Prior to Sgt. Pepper, the one and only Billy Shears, the lyrics to the music I listened to and would play on the radio, was laced with the innocence of the times. The nation was between wars, there really wasn't much to protest, and life was good. Music was about love, girls named Carol, and Purple People Eaters. Pegged pants, madras shirts, and Thom McCann pointy-toed shoes became my uniform. As naive high school kids in the early 'sixties, we sat in cars watching B-rated horror movies, and washing down Chicken In The Basket with shakes at drive-in restaurants.

Driving a car was the springboard to freedom from home rule. On my sixteenth birthday, I got my driver's license and drove the family Nash Rambler station wagon without an adult in the car for the first time.

My first stop was two blocks down the street to pick up my best friend, Billy Thompson, who was only fifteen at the time. We cruised aimlessly, listening to WING blaring out of the tinny sounding 3-inch speaker in the dashboard. The best part of driving was knowing that from that day on, I would never have to take a bus to school ever again. No more baloney and cheese sandwiches on white bread from a brown paper bag at lunch for me. I could now drive to the Vic Cassano and Mom Donisi Pizza Parlor a few miles away for a Vesuvius Steak Sandwich topped with brown gravy and banana peppers on an Italian sweet roll.

I started hanging around the WING studios in downtown Dayton whenever Bob Holliday was on the air. I swept the floors, filled the Coke machine, and functioned as a gopher for the jocks. My duties included schlepping hamburgers from Frisch's Big Boy across the street for Lou Swanson, Dave Parks, Gene "By Golly" Berry, Skinny Bobby Harper, and Mike Nardone (who coined the phrase *help stamp out daylight,* and wore dark sunglasses twenty-four hours a day). Anything to be inside that radio station.

Then in 1962, Bob said, "WING is going to participate in a Junior Achievement program. We're turning over the air waves for one hour, every Sunday night, to a group of high school kids. They'll have their own air staff and advertising sales department. Would you like to be one of those kids?"

Of course, I said, "Yes" and got to be the DJ once a month. Gary Sandy, who later appeared as the PD (program director) on WKRP in Cincinnati, was also one of us, as well as Paul Iams, whose father founded Iams Dog Food. This was my first actual on-the-air experience in radio, and I got a great deal of satisfaction knowing that the kids at school were listening to me. Thanks to Bob, I was eventually hired part-time to run the board now and then for the jocks. Since he was the PD of the station and my mentor and role model, the seeds were planted to have a career in radio.

I also became Bob's roadie for his lucrative side business of producing record hops throughout Montgomery County. Every weekend, he was hired by area high schools and for nonprofit organizations to supply the music for dancing. Someone had to cart all that heavy equipment to these events. That was me. Eventually, Bob's hop business became so big that he bought a powder blue 1959 Ford Fairlane station wagon for me to transport the sound system throughout the county. Painted along both sides of the car in big letters were the words: *Holliday Hop Caravan* and *WING Radio*. In small lettering, just below the radio antenna was painted, *Ron Copeland, Engineer*. I *loved* driving that car to school.

In the summer of 1962, every Friday night the Holliday Hop Caravan would set up shop at the LeSourdsville Lake Amusement Park's outdoor ballroom just south of Dayton in Hamilton, Ohio. Literally hundreds of high school kids attended these shows to see artists like the Beach Boys, Bobby Vee, Johnny Tillotson, Ray Stevens, and Gene Pitney performing live. One of the regular bands was called the Rick Z Combo. They were three twelve-and thirteen-year-old kids from Union City, Ohio. Rick played lead guitar, his brother Randy was the drummer, and the base player was about the same age. Three years later they became the McCoys and had a number one hit with "Hang On, Sloopy." Around 1970, Rick changed his name from Zerringer to Derringer. A few years later, he was laying down incredible guitar riffs on "Frankenstein" for the Edgar Winter Group.

In 1963, the teenage Beach Boys appeared in the ballroom, and the place was packed. At this point in their career, their hits were "Surfin Safari," "Surfin USA," "Surfer Girl," and "Little Deuce Coup." It was a hot, humid August night, and with the smell of corn dogs and cotton candy in the air, the

Beach Boys got all us land-locked Ohio kids wishing we could hit the surf on a California beach.

"After this next record ends, bring up my mic," Bob instructed me. "When I introduce them, hit 'em with the red spotlights."

The record started fading out, and the hum of the speakers at full volume could be heard in the absence of the record. Bob stepped up to the microphone and said, "Direct from Los Angeles. C'mon kids, let's hear it for the Beach Boys!"

As the guys took the two steps up to the small stage and headed for their instruments, the excitement in the air was intense. Carl Wilson started tuning his guitar, and his brother Dennis took his seat behind the drums. A few seconds in to his warm up, all of a sudden, he jumped off his seat and yelled out, "Hold on, I gotta take a shit." He then bolted off the stage and disappeared into the connecting dressing room. I don't think the audience could hear him say that, but Bob and I sure did.

I started another record just to cover up the nervous silence, and Bob assured the restless audience, "It'll just be a few more minutes, everybody hold on."

Dennis came bounding back on to the stage and the show went off without a hitch. What a night it was. These were exciting times for a former wall flower like me. A senior in high school, and I was actually standing on the same stage with some of the biggest stars of the day.

I graduated from high school in June of 1963. One day, I was shooting hoops at a friend's house when we started talking about girls.

"Do you remember that sophomore Beth Castle?" I asked. "I'd love to go out with her."

"Well, why don't you call her up and ask her?" he said.

"Uh, well, she probably doesn't even know who I am, and besides, she'd just say no."

"C'mon chicken, call her up."

"You're right, I'm chicken," I said. "Okay, I'm going to do it. I'll call her. In a day or so."

✤

"Hello, Mrs. Castle, is Beth there?"

"Who is this?"

"Uh, it's, uh, Ron Copeland from high school."

"Beth, there's some guy named Ron Copeland on the phone for you."

"Hello."

"Uh, hi, Beth, uh, this is, uh, Ron Copeland from school. You may not, uh, know me, but I was just wondering if, well, if you'd like to go out with me some time."

"I know who you are," she said, "but I'm grounded. My dad won't let me date for a while."

"Well, how about if I call you again some other time?"

"Call me back tomorrow. I'll talk to my mom."

"Okay, I'll call you tomorrow."

To me, Beth was the sexiest, most beautiful girl at Beavercreek High School. In reality, we were in two very different worlds at Beavercreek. I was in the popular, crew-cut, squeaky clean category, while Beth was in the rebellious, greaser crowd. In spite of that, my urges over powered any hesitation I might have had to pursue her. All I knew was I had to have her. I almost didn't call her back. But I did. She had talked her mom into talking to her dad and had assured them both that I was one of the "good boys" and didn't have a bad reputation like some of the guys she had dated. By convincing her parents that I was okay and agreeing to go out with me, she was able to get "un-grounded," which I'm sure was part of her motivation to say yes to a date.

For the rest of that year, we went everywhere together.

In January of 1964, I was madly in love, but I had promised my dad that I'd go to college. I spent all my savings on one semester at Ohio State, which was fifty-two miles away in Columbus. Every Sunday night, I'd tear myself away from Beth, jump into my 1951 Plymouth sedan, and drive back to school. All week long, I'd mope around the dorm and reluctantly attend classes. What I knew was that I wanted to be in radio and no one in any college could ever teach me more about that than I could learn by actually working at a radio station.

The most difficult Sunday night of that time of my life was on February 9, 1964. That was the date the Beatles first appeared on the Ed Sullivan Show. At this point, long before Beatle songs about *tangerine dreams and marmalade skies* were written, it was all about holding hands and "P.S I Love You." The Beatles' influence on our thoughts and our appearance was dramatic. Nehru jacket, Beatle haircut, miniskirts, and hair extensions. The hormones were raging, and as we sat, young and in love, on Beth's parents' living room's hardwood floor only a few feet away from the black and white TV screen, thoughts of the future were intoxicating even without drugs or alcohol. With the help of John, Paul, George, and Ringo, and my association with Top 40 Radio, we were on the cutting edge of cool. When that show ended, and in the dead of an Ohio winter night, I had to say goodbye to the love of my life and drive to a place I hated to be, I made a decision. I was going to finish that semester and quit school. Working at WING and being with Beth were all that mattered.

My mother was very liberal for the times. She took me aside one day in late 1963 and said, "Ronnie, I know you and

Beth are having sex. Think about how bad it would be if she got pregnant. Here, take this, and please start using it."

Mom gave me a tube of EMCO foam, which was used for birth control back then, since the pill hadn't yet been invented. Great. No more condoms. No more pulling out. Fire away. All systems go. Guess what? That shit doesn't work.

I never considered not staying with Beth and raising our child together.

One big problem. We weren't married yet. And then there was that huge problem of Beth's father killing me. So we both decided we'd go to our mothers, who were more understanding than our dads. A plan was hatched. I don't remember which one of us found out about this justice of the peace in Covington, Kentucky, who was known for taking bribes. All we had to do was slip him an extra $20, and he'd pre-date the marriage certificate to make it look like we had secretly been married for six months already. That way, it would be okay in the eyes of society, and more important, Beth's dad wouldn't kill me.

We headed for Covington, Kentucky, located just over the Ohio River across from Cincinnati, in Beth's dad's 1962 Dodge station wagon with her mom at the wheel.

The shotgun wedding went off without a hitch, and with marriage certificate in hand, we headed back to Dayton to face the music. Our dads both accepted the fact that, A, we were married, and B, we were pregnant. No problem. About a year later, after the baby was born and my dad got his very first grandchild, he took me aside and said, "Son, you didn't think I was stupid enough to believe that, did you?"

Teddy & The Rough Riders with
me and best friend Billy Thompson
in back

Bobby Vee back up by Teddy &
The Rough Riders at Lesourdes-
ville Lake, circa 1962

Bob's Beach Boys autographs
that night. Why I didn't get one,
I'll never know.

Paul Iams and Gary Sandy during
Junior Achievement Show

Me dressed like The Beatles,
Beth dressed like
The Shangra Las

Jocking for Junior Achievement
on WING

Hype sheet for the Holliday Hop Caravan

CHAPTER 3

"EIGHT DAYS A WEEK"

In 1964, after apprenticing at my hometown radio station since the eighth grade, I was ready to go on the air full time and make a living with my mouth. The problem was, my voice had not completely matured yet, and quite frankly, I stunk.

Walt Turner, one of the DJs at WING said, "Boy, if you want to be a DJ at WING, you need to pay some dues. I started out at a little radio station in Cynthiana, Kentucky, and I'm pretty sure I can get you a job there. You need to get in the trenches and just do it for a few years. It's farm country, but announcing hog futures would give you some polish."

He got me hired, and in June of that year, at age nineteen, I grabbed my pregnant eighteen-year-old wife, and we drove to Cynthiana in our ugly brown 1953 Chevy Bel Air with headlights that worked now and then. The problem was, they picked the wrong time to stop working, like just as the sun was setting. But it got us there. Since my starting pay was a buck an hour, commonly referred to in radio as a dollar a holler, and I had a child on the way, where to live would be a problem. Laurence McGill, the kindly southern gentleman who owned WCYN, also owned a real honest-to-God redneck, trashy trailer park. He allowed us to live in a tiny 1950s-style trailer for free. It was so small that Beth and her growing tummy almost had to back in. It was home, and we were too young and in love for it to bother us in any way.

Within a few months, we had enough money to move in to a small, one-bedroom apartment on the second floor of the A&P Supermarket. It was right next to the OK Used Car lot and across the street from the front door of the radio station.

Since my air shift started at 6 a.m. and ended each day at 6 p.m., I wanted to live close to work in case my headlights failed. My job was to sign the station on every morning, and over-sleeping was not an option. Tens of farmers milking their cows and slopping their hogs at 6 a.m. expected me to be on the air telling them the price of pork bellies and grain futures. I took this responsibility seriously. One morning, I bolted upright in bed, looked at the clock and saw that it was five minutes till sign-on time. I threw my clothes on, mad-dashed across the street, bounded up the two flights of stairs, and began the process of turning on the transmitter.

The mega-sized glass vacuum tubes inside took a while to warm up, but by 6:05, I had that puppy up and running as the national news feed was being broadcast from ABC.

Now I could relax and go about my morning sign-on duties. First was tearing down the wire. The Associated Press teletype machine had been printing reams and reams all night long, and before I could read the state and local news, the long rolls of paper splayed all over the floor had to be ripped apart and categorized for easy reading on the air.

As I did this, I said to myself, "Strange, usually the weather forecast has been sent by now. Oh well."

I got the coffee going, turned on all the lights, and by fifteen past the hour, I sat down and opened the microphone.

"Good morning everyone, this is Ron Copeland on WCYN radio. The time is..." Then, a long silent pause as I finally truly woke up, looked closer at the clock, and painfully realized it was 3:15 a.m., not 6:15 a.m. I was on the air all right, but three hours before sign-on. My conscientiousness bit me in the ass. I never said another word, realizing that at that hour of the morning, it didn't matter. Nobody was listening anyway. I just turned off the microphone and the transmitter, went down the stairs, crossed the street, walked up the stairs with the familiar scent of the grocery store wafting through my nostrils, jumped back in bed, and went to sleep.

But before I did, I nudged Beth and said, "You won't believe what just happened."

I didn't actually work twelve straight hours each day. Monday through Friday from 6 to 9 a.m., I did the Morning Show, a mix of the afore-mentioned hog reports, news, weather, and country music. When I got off the air, my duties were to type up the news headlines of the day on station letterhead, run off two hundred copies on one of those old mimeograph machines where you crank the handle and got that totally unique aroma coming off the paper. Then, I'd hike my ass from business to business in downtown Cynthiana, a four-block area, and drop off a stack of news headlines on each business counter for their customers.

I'd then go home, take a quick nap, and be back on the air at 11:30 a.m. to conduct the WCYN Trading Post. It was a one-hour show where the local yokels would call in and offer goods for sale, like a flea market on the air waves. Calls like, *"This here's Estel Sipples and ah have fourteen fence posts in*

good condition fur sale or trade. If you been looking fur some
good, used fence posts, call me at Adams-3440. Thank yew. "
Or, *"Hi, this is Sudie McGladdery, reminding all ya'll*
that tonight is bingo night at the Main Street Southern Baptist
church. Bingo will be followed by choir practice and butter-
milk pie for dessert."
Being nineteen years old and freshly turned onto the Beatles, I couldn't get very excited about this show. At 12:30 p.m.
when it was over, I'd go back home and count the roaches that
worked their way up from the produce department below. And
then, it was back to the radio station for the Afternoon Drive
show from 3 to 6 p.m., or whatever time sunset was.

In a sleepy little farming community like Cynthiana, Kentucky, smack dab in the middle of the Bible Belt, with a population of only six thousand, the one and only radio station in
town did not qualify for free records from the promoters. Any
current hits of the day played on WCYN had to be paid for in
cold hard cash. Since I thought I was the only one within fifty
miles who was "hip," I had to scour the Billboard charts for hit
records for my afternoon show.

Someone had to turn those good country folks on to "real" music, so kindly old Mr. McGill said, "Son, I think that
kind of music will send you straight to hell, but we need listeners, and if you think this will do it, here's a five dollar bill.
Go buy some records."

Every Saturday, my day off, I'd drive the forty-five miles
to Lexington, to buy five new 89-cent 45 RPM records at the
one and only record store in town. I was really into the music
and anguished over which ones to buy and which ones to wait
a week to buy. Because of my faulty headlights, I'd leave Lexington in plenty of time to be home before sunset.

I couldn't wait to get back on the air to turn central Kentucky onto the British Invasion every day. But, alas, the half

hour before sign-off every night, the owner insisted I change the format to the Sammy Kaye Serenade. Mr. McGill just loved 1940s-style big band music, and well, after all, he was the owner. He had a 78 RPM record collection of every Sammy Kaye LP ever sold.

While living in the trashy trailer park, Beth and I made the decision that when it came time for the baby to be born, the actual birth would take place at the same hospital I was born in—Miami Valley Hospital in Dayton. That meant that during the last few weeks, Beth would have to stay at her mom's while I continued working in Kentucky. Then, the minute it looked like the baby was going to be born, I'd jump in the old rattle-trap Chevy and hightail it to Dayton. When the call came, I bolted out and headed to Highway 75 North and drove as fast as the car would go.

A few blocks before I got to the Castle home, during that twilight zone between dusk and dark, it happened. No headlights. I was at an intersection in a somewhat rural area. Thank God there were no other cars around. I strained my eyes trying to keep the car on the pavement. As I was making a left turn, I missed completing it by a few feet, and as in a dream, I was no longer riding in an old car. I was atop a bucking bronco careening over a drainage ditch in to a freshly plowed field. Fortunately, I hadn't been going very fast, so I basically just coasted to a halt. After a few minutes of trying to digest what had just happened, I started the engine, and very carefully drove out of that field and completed the last few blocks in the dark without further incident.

When I arrived, Beth and her mom were already at the hospital. Here I was in a car with Beth's dad, a grizzled old World War II vet who worked for the city of Beavercreek.

"You didn't think I was dumb enough to believe that cock and bull story about you two being married earlier, did you?"

Gulp. I guess he had a right to be suspicious.

When we got to the hospital, the baby had already arrived. I entered the room to give my wife a big hug. The baby was in a bassinet next to her, covered with a blanket. Not the way it's done in the movies, I thought. I approached the bassinet fully anticipating my first meeting with my new son, Ronald Thomas Copeland Junior. My dad had fathered five boys, no girls. Of course, I would have a boy.

Uh oh, where's the penis?

To my delighted surprise, I had fathered a beautiful little girl, and Ronald Thomas became Rhonda Ann.

Knowing we were going to soon outgrow the dumpy trailer we had been living in, before driving to Dayton for the birth of the baby, we had rented the apartment above the grocery store, but there was one problem. No furniture. Ah-ha! Uncle Jack to the rescue. Three full rooms of furniture for only $275 dollars. With payments of $10 dollars a month, we'd have it paid for in about two and a half years.

On Sundays, I was on the air from 10 a.m. to 8 p.m., a straight ten-hour, on air shift. After Rhonda was born, Beth would set up camp in the studio with me, and we'd play cards during records. Halfway through the shift, she would go downstairs, around the corner to the food counter at Begley's Walgreen Agency Drug Store to retrieve a couple of grilled cheese sandwiches and a coke to hold us over till dinnertime.

After a year, I left Cynthiana, armed with the knowledge that I never wanted to have to read another hog report as long as I lived. Life was good. I had a family, a career, and was ready to climb the ladder of success in radio.

Our apartment on the second floor of the A&P Grocery Store next to the OK Used Car Lot. It was across the street from WCYN.

On the air at WCYN

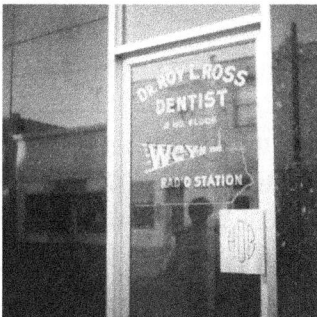

Dentist on the second floor, radio station on the third floor

CHAPTER 4

"SATISFACTION"

At WCYN, I was working approximately sixty hours a week but paid for only forty of them. At a dollar an hour, that was just not enough to support a wife and a baby, even in 1964. The only way I would ever break through to the next pay level was to acquire my First Class License. Back in those days, the FCC required radio stations to have a first-class-licensed engineer in the building at all times. During regular business hours, the larger stations had a full-time engineer on duty to keep the equipment operating and in good repair. Savvy station owners soon figured out they could save big bucks by hiring jocks who also had a First Class License, therefore, eliminating that second paycheck. Schools like the Elkins Institute of Technology in Chicago, where I went, and the famous William B. Ogden Radio School in LA began popping up. They simply taught a jock how to pass the test, not how to truly understand electronics.

It was 1965, and my father, step-mother, and younger brothers had moved to Rockford, Illinois, as my dad chased yet another job opportunity. He was my biggest fan. Years after I had made it to the major markets, he told me of a chance meeting with a guy named Jim Smith, who was the program director of WING in Dayton just before Bob Holliday replaced him.

"Mr. Copeland, can I talk to you for a minute?"

"Sure, Jim, I have a second," my dad replied.

"I've heard some of your son's voice work, and you need to steer him in another direction. He just doesn't have what it takes to make it in radio."

"What do you mean?" my dad asked.

"He doesn't have the voice for it."

"Well, I think he has a really good voice. I'm not going to tell him that."

"You'll save him a lot of heartache and disappointment down the road if you tell him."

"Thanks, Jim, but no thanks."

And that was the name of that tune.

I am so grateful he saved that bit of information because I was already insecure enough. I didn't need to hear that, and my dad knew it. Instead of passing along the mean-spirited criticism, he offered to pay the fee for me to attend the Elkins Institute of Technology in Chicago. He discovered this First Class License loophole on his own. Unbeknownst to me, Dad had met with the owner of Rockford Top 40 radio station WJRL. He wanted his son, daughter-in-law, and granddaughter to be closer to him, so he took one of the reel-to-reel air checks I sent him now and then to the meeting to try to get me hired.

The station owner was John Rogers Livingston; hence, WJRL. He was an old time, cigar-chompin' radio station owner who resembled Edward G. Robinson.

As Dad walked in to the office to take a stab at getting me hired, the stench of stale cigar smoke filled his lungs. Peering through the thick cloud, he said, "Mr. Livingston, my name is Clyde Copeland, and my son is a disc jockey in Kentucky. Could I get you to listen to this tape of him? He'd love to come to work for WJRL."

"Well, okay. We do need a disc jockey for the evening shift," he grunted in a raspy, impatient voice. After listening to a few minutes of the tape he said, "Well, he needs a lot of direction, but I guess my program director could work with him. Does he have a First Class license? I couldn't possibly consider this kid without one."

"Well, no he doesn't," Dad admitted.

"Then come back when he gets his license, and I'll hire him. Now, I have another meeting. Thanks for stopping by," he said, still clenching the cigar between his teeth as he talked.

For six weeks, I was separated from my wife and daughter as I attended Elkins. With little money, I spent those weeks living at the YMCA flea bag hotel in downtown Chicago. I hated it. In school I never learned a thing except how to pass the test, which I did. At a hefty pay increase from a $1.00 to $1.25 per hour, I had arrived. The first song I played at WJRL was "Satisfaction" by the Rolling Stones, summing up exactly how I felt about getting that license.

WJRL turned out to be a really shitty place to work. As I did throughout most of my early days in radio, I always scooped up the weekly issue of *Broadcasting Magazine*, which had job openings listed on the last few pages. I constantly replied to those ads by tossing a tape and a resume in the mail. Six months after being hired at WJRL, I got a nibble. I was offered morning drive at WDUX Radio in Waupaca, Wisconsin, about a hundred miles north of Rockford. I took it.

Beth and I shared the same feeling of adventure when it came to picking up and moving a thousand miles at the drop of a hat. Kids, cats, a truck load of furniture, and a pay increase of only ten bucks a month. It didn't matter. It was the call of the wild and wooly.

The most important thing that happened there was the birth of my wonderful daughter, Robyn Lynn. Then, after six months in a cold Wisconsin winter, WJRL Radio was purchased by a professional broadcast group who came in, changed the call letters to WYFE, and hired me back.

Rockford is only ninety miles from Chicago and was the home of legendary radio station WLS. In 1965, listening to Clark Weber, Ron Riley, Dick Biondi, Dex Card, and Art Roberts was like going to disc-jockey school for me.

I was put on afternoon drive at WYFE and decided to call myself Rik O'Shea. I changed my name to distance myself from WJRL, where I had used my real name just six months earlier. There was a comic strip called Rik O'Shea back then, and I always loved the double entendre. Another radio name I liked that had two meanings was Justin Case, which I never got around to using.

At WYFE, all the training I received from Bob Holliday as the roadie for the Holliday Hop Caravan started paying off. I did record hops for extra money, put on Battle of the Bands events at high schools, and a few times, brought in big groups like Paul Revere & The Raiders, Eric Burdon and The Animals, and The Yardbirds for concerts at the Rock River Roller Palace. There I was, standing on the stage with Jimmy Page and Jeff Beck, who were in the Yardbirds at the time. Beck bumped his head on a low hanging water pipe that was near the makeshift stage. In 1966, I had no idea who Jimmy Page was yet, since Led Zeppelin was still a few years away.

During this period in my career, I also managed a band of high school kids called The Jacemen. Bob Holliday was always managing bands when I worked for him. I was learning. They'd join me at high school record hops and usually won in the Battle of the Bands competitions.

Battle of the Bands competitor

One of the many high school bands who competed at my Bands Battle of the Bands at the Palace

WYFE survey where I went by the name of Rik O'Shea

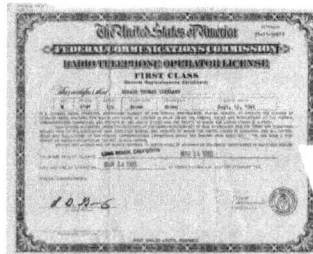

My First Class License

Mark Lindsay of Paul Revere &
The Raiders appearing at the
Rock River Roller Palace in 1966

Tony, lead singer of the Jacemen

CHAPTER 5

"CALIFORNIA DREAMING"

Toward the end of 1966 at WYFE, restlessness was set-
tling in. Since my days at WING, I had always stayed
in touch with Bob Holliday. By this time, Bob, tired of
Dayton and the snowy weather, took on the job of PD at a
10,000 watt AM powerhouse in Tucson, Arizona. The call letters were KTKT, and it was known as the 10,000-watt flower
pot, in honor of the new hippie movement. Bob knew that I,
too, was sick of Midwest weather and the lure of the wild,
wild, west might appeal to me. He was in need of a nine-to-
midnight jock. Bob knew my work ethic and strong desire to
make good radio. He could only pay me a few dollars a week
more than I was making in Rockford, but it wasn't about the
money. It was the adrenaline rush of a new experience, and
this time I could escape the dismal Midwest where my short
radio career had been centered thus far.

From the time I heard "California Dreaming" by the Mamas and Papas, I knew that someday, I'd make it to the west coast, and Arizona was a hell of a lot closer to the Pacific Ocean than Illinois. I didn't think twice about his offer. Are you kidding me? A much bigger city, a signal that could be heard a thousand miles away, and above all, another opportunity to see the USA in our 1965 Mustang. At the ripe old age of twenty-one, off I went to a city that the Joe Bonnano Crime Family called home. A place with rattlesnakes, cactus, scorpions, intense heat, and the land of the most beautiful skyscapes you could ever imagine.

As always, this called for another name change. I became Ron Knight. I can't recall how that came about, but I can tell you it's not easy to switch to a new name and remember to use it every time you turn on the microphone. More than once, I'd accidentally bark out Ron Copeland or, Rik O'Shea. This problem went away quickly but required a lot of concentration in the beginning. In February of 1967, I became an official member of the KTKT *Swinging Seven* as the jock line-up was referred to. That line up was Jerry Stowe, 6 to 9 a.m., Dan Gates, 9 a.m. to noon, Bob Holliday, noon to 3 p.m., Frank Kalil, 3 to 6 p.m., Joe Bailey, 6 to 9 p.m., me, Ron Knight, nine to midnight, and Lee Poole, midnight to 6 a.m.

Until then, I was still drug- and alcohol-free, having never tried either in my twenty-one years of living. One night, after getting off the air at midnight, I jumped on my little Honda 50 cc putt-putt motorcycle and headed for home. We had rented a small house on the side of a hill a few miles out of town on the western edge of Tucson. It was a crystal clear, warm summer night, and as I motored along at the screaming top speed of a tire-burning 35 miles per hour, I could see for miles and miles in any direction. I was the only vehicle anywhere in sight, all by myself in the desert during a bright full moon of an evening.

Way off in the distance to my left, I could see what seemed like a large, disc-shaped object hovering a couple of hundred yards above the ground. It was slowly working its way toward me, and as it meandered my way, every now and then, an intensely bright spotlight would illuminate the ground directly under this flying saucer. As it came closer and closer, apparently headed on a course that would soon intersect with mine, I mentally tried to will the motorcycle to go faster.

This thing was totally silent while in motion and only made some sort of whooshing noise when the light pierced the semi-darkness below it. Apparently, I was nothing of interest to whatever it was, and it slowly worked its way past me to the right and, in the course of time, became a distant spot on the horizon. Life west of the Mississippi was taking me into a whole new level of awareness. Papa had a brand new bag.

Beth and I were huge James Brown fans, and when we heard he was coming to Tucson, we were going to be there come hell or high water. Fortunately, the concert promoter advertised on KTKT, which meant we could get in free. In the early 1960s, most of Brown's records were only hits on hard core R&B radio stations. But in 1965, all that changed because of two mainstream top ten hits, "Papa's Got a Brand New Bag" and "I Got You (I Feel Good)." We were James Browns fans way before that, back in Ohio, thanks to Bob Holliday's attempts to break his records on WING years earlier. He was a big supporter of James Brown and was playing "Night Train" in heavy rotation as early as 1961.

We were fired up about seeing him live. This was not a typical James Brown concert where he was playing in an auditorium to a huge crowd. Not in those days. The show took place in a high school gymnasium. He and his band were positioned on a simple two-foot high riser on the gym floor. The audience stood right in front of him, as there were no chairs. Beth and I were two of the few white fans in the crowd. In 1967, spearheaded by the Beatles' "Sgt. Pepper" release, music took on an entirely new direction. The Doors wanted to "Light My Fire," and Jimi Hendrix wanted to know if I was experienced. Having been raised in a religious family and married to the first girl I ever had sex with, I was the least experienced jock I knew. Slowly but surely, out-of-control urges began to replace common sense. I knew I would give in eventually. Temptation was everywhere I looked, from the braless receptionist to the miniskirted hit line chicks. I wanted desperately to enter the world of sex, drugs, and rock & roll. The only problem was, I was married. I decided I needed a temporary separation. I wanted to make it legal and avoid going to hell for committing adultery. It made perfect sense to me at the time.

"Mr. Lawyer," I said. "I would like to hire you to draw up the papers for a temporary, but legal separation from my wife."

"I'm not sure what you mean. Tell me, what is it you are trying to accomplish here?" he asked.

"Well, I uh, just want a little break from my marriage but I want to, well you know, make it legal." I sputtered.

"Make it legal? What does that mean?"

"You know, make it legal. Legal in the eyes of God."

"So you can, what?"

"So I can, I don't know what I'm trying to say, I just…"

"Mr. Copeland," he said. "No one is going to put you in jail for cheating on your wife. As far as God goes, that's between you and him. I can't help you there."

So much for that idea.

As a disc jockey, you can start believing your own press releases when you answer the request line, and a sexy-sounding female indicates she wants to jump your bones. Ask anyone who was a DJ on a Top 40 radio station back then, and he'll tell you the same thing. There it was. That fire-breathing dragon called temptation in my face again. Beth could feel the tension coming out of my pores, and I'm sure it made her angry. She retaliated by getting a job at the Red Dog Saloon, a bar and restaurant. It was located inside Old Tucson, the site for more than 300 film and television projects since 1939, including *Gunfight at the OK Corral* with Burt Lancaster, John Wayne's *Rio Bravo*, and Jimmy Stewart's *Winchester 73*.

We always practiced *I'll show you* throughout our marriage. In the 'sixties, Old Tucson was a tourist destination with re-enacted gunfights in the streets along with other western movie-themed attractions.

Beth had to dress in the attire of the late 1800s, and she was looking very sexy to me, as well as the small time cowboy actors who performed in the re-enactments. Now it was time for me to feel insecure and jealous. All of this led to a complete emotional showdown between us, preceded by days and days of not speaking to one another.

I suggested she take the kids and go live with her mother and father in Dayton for a while to give us some space. Sometime in the middle of the summer of 1967, we separated, not legally of course, and I took my first steps into a world I had only dreamed about.

The first night after I drove my family to the airport, I played cards with some coworkers and proceeded to suck down a whole quart of cheap gin straight out of the bottle. Anyone who drinks knows that's a ticket to the swirling white bowl. That led to my first hangover. Not long after, it was *cannabis sativa* time. The first time I smoked marijuana, I was alone in my apartment listening to "Purple Haze" by Jimi Hendrix. The walls were all wavy, the ceiling became the floor, the floor became the ceiling, and life was never the same.

An attractive waitress who worked in a restaurant we all hung out at became my first of many one-night stands. Then, there were those hit line chicks. I was young but high school girls were still underage. That didn't stop the other jocks and ultimately, it didn't stop me, either.

I was on the air one Sunday night and had played a new song by the Stone Poneys. The song was "Different Drum" written by Michael Nesmith of The Monkees, and the Stone Poneys lead singer was Tucson native Linda Ronstadt. Bob Holliday had gotten an advance copy of the record through his connections in California before it was released. Since KTKT was Linda's hometown radio station, it was important that we got to break the record first. About a half hour after playing it, the request line phone rang.

"KTKT hit line. What can I play for you?"

"Is this Ron Knight?" the sexy female voice asked.

"Yes, it is. What would you like to hear?"

"Hi Ron, this is Linda Ronstadt, and I wanted to thank you for playing my record. I just drove home from the recording studio in Los Angeles today and was happy to hear it being played only a couple of days after we recorded it. Thank you."

Cool. I almost asked her for a date but was too shy. Later, I heard she had dated some of the jocks at KTKT, so I really kicked myself for not asking. Considering my luck with women in those days, it was probably just as well.

Somewhere along the way, I met Dino Day, the jock who was on the air at the competing top 40 station, KIKX. We joked about our names, Knight and Day, and we became great friends. We were aware that the audience who listened to our radio stations were not loyal in any way. They would bounce back and forth, listening to both stations. Late one Sunday night, out of boredom, we came up with an idea that would make our listeners question their sanity. After 9 p.m., most radio stations would loosen up the playlist and let the jocks experiment with harder rock that you'd never hear in the daytime. We were permitted to play the long versions of songs that had been edited down for daytime airplay. This particular night, we decided we would both play the eighteen-minute long version of "In-a-Gadda-Da-Vida" by Iron Butterfly, and would start the song at exactly the same time. For the listener punching back and forth on the dial, it had to be a real head scratcher. Our program directors never had a clue about our antics since it was 3 o'clock in the morning, and our bosses were sawing logs.

Dino and I would drive to Mount Lemon, and sometimes to the top of A Mountain and smoke dope. "A" Mountain is a small protrusion that overlooks downtown Tucson. From maybe a thousand feet up, we'd get stoned and watch the twinkling lights of the city.

One night, I was sitting around the production room with Dino and several other jocks dreaming about the day we'd make it big in radio. We were talking about how to improve our on-air sound when Dino said, "A friend of mine in Los Angeles sent me this air check last week. It's a tape of some guy on a station in Hollywood."

"California? Really. Let's listen to it."

What I heard on that air check created a drastic change in what I sounded like on the next night's air shift and for years to come. My energy level quadrupled. My pace quickened, and I was not the same. What I heard that night had the same effect on me that it had on hundreds of other disc jockeys across the nation. Like me, they were exposed to The Real Don Steele, the legendary and innovative disc jockey on KHJ Los Angeles. He re-wrote the definition of disc jockey. His rapid-fire delivery, his humor, and the back-to-back hits he played made him the most emulated DJ in the country.

Don Steele was able to concentrate on his sound because he, like all of the major market jocks, had an engineer. At KTKT, I wasn't as fortunate.

I always hated the technical side of radio. While I was on the air, additional duties included taking meter readings and monitoring the transmitter, since I had a First Class License. KTKT had two transmitters. One was a 10,000-watt monster that glowed while it hummed and whirred and vibrated. The other one was much smaller and only 1,000 watts. Like most radio stations at sunset, we were required to lower the power of the signal. This meant physically switching from one transmitter to the other.

One night, I put on a long record and trudged in to the transmitter room behind the studio. The two transmitters were side by side. After throwing the switch on the 10,000-watt transmitter to off, I shoved the 1,000 watt transmitter to on. Sparks flew, and smoke wafted through the air, and then, silence. When I threw the switch on the 10,000-watt behemoth back to on, just to stay on the air, it blew up too. More sparks. The smell of creosote filled the air. I stood there leaching out fear, realizing I had just blown up both transmitters, and we were off the air. Since this nightmare was taking place after regular business hours, and I was the only one in the building, it was totally quiet. No rock and roll music spilling out of the studio speakers. Nothing.

Huddy, our fuddy-duddy chief engineer, was in his seventies and had been at KTKT since radio was invented. These transmitters were his babies. He was a real 1960s nerd with a pocket protector and a two-day growth of beard. He'd burp, fart, and generally speaking, smell bad. Huddy had very little patience for a young whippersnapper who hated his transmitters.

"Hello, Huddy, this is Ron down here at the station. We are off the air. Both transmitters are not working. I did everything I do every other night but, well, something's gone wrong, and you better get down here now."

"Well, shit! Goddam! What the hell did you do? Oh hellfire! I'm on my way." As the phone was being put back on the receiver, I heard him uttering cuss words I'd never heard before. Fortunately for me, I had done nothing wrong. It was an absolute coincidence that both transmitters decided to fail at the same time. We were off the air for three days, waiting for parts to arrive. Don't look at me. I didn't do it.

Toward the end of 1967, I started missing Beth and the kids big time. I compared the thrill of being single with being married and made an easy choice. The single life sucked, and I wanted to get back together with my family. Out came the *Broadcasting Magazine* classified ads again, and out went the air checks and resumes to stations all over the country. After many hours on the phone with Beth saying how sorry I was for what I had done to her, we both agreed it was time to get back together.

She knew I was trying to get a job in Ohio, but when I landed nine to midnight on WOHO, Toledo, I never told her. Instead, I arranged to surprise her with the help of my mother-in-law. She was very happy to hear we were reuniting.

I accepted the job at WOHO. Around December 22nd, I got in my 1965 Mustang and headed for Dayton. My plan was to knock on Beth's front door on Christmas Eve. I also brought her a present, an ounce of grass. I wanted to be the first person to turn her on.

I decided to drive at night to avoid traffic and sleep in the daytime on my way back to Dayton. The first night, at around 3 a.m., I was barreling down Texas Interstate Highway 10, headed to Odessa, for a stop off to see my Uncle Houston and maybe crash on his sofa for a bit. About an hour across the Texas state line, red lights started flashing behind me, and the piercing sound of a siren slowed me down to a complete stop.

"License and registration please. You were pulled over for speeding. You were doing 75 in a 65 mile an hour zone," said the intimidating Texas highway patrolman. As I reached for the glove compartment to get the registration, fear shot through me like a bolt of Arizona monsoon lightening. It hit me that the clear baggy filled with pot was sitting right next to my registration, and his flashlight was following my hand to the glove box. Holy shit.

Luck was with me. He never saw nor smelled it.

I have always made a point to be overly courteous when dealing with an officer of the law. Never more so at that moment.

"Gee officer, I looked at my speedometer not long before you pulled me over, and I swear I was only doing 65." I prayed my Beatle haircut would not piss this guy off.

"Well, my radar said 75…wait a minute." He shined his light on my rear tires.

"You know, those tires look larger than the tires that came with this car."

"I know, officer. Before I left Tucson to drive to Dayton to be with my wife and two little girls for Christmas, I had to buy tires, and those were the only size of used ones I could afford."

"Well that's your problem," he said. "Oversize tires cause the speedometer to show less speed than you're actually doing.

I'm going to let you go this time, but remember this as you continue your drive."

"Thanks officer. I really appreciate this, and I will watch my speed much closer from now on."

The rest of the drive was uneventful, and on Christmas Eve 1967, I pulled into the driveway on Turnbull Drive in Beavercreek, Ohio, ready to re-start my married life. Beth was totally surprised. Christmas that year was an emotional family reunion. We were back together again and extremely happy.

My first day at WOHO in Toledo was New Year's Day, 1968. I was filled with excitement when I played "Dance to the Music" by newcomer Sly Stone, a wild and crazy Oakland, California disc jockey. In a soundproof studio and alone in the building, I cranked the volume up to maximum level and let those pear-shaped decibels wash over me with the feeling of euphoria.

I knew that I was sharing this new electrically charged anthem with hundreds of other people on the first day of a new year. In spite of the Watts riots that bled over to many other major American cities, the political assassinations, and the Viet Nam War, Sly & the Family Stone's over-the-top energized rock and roll dance tune was indicative of the times. The let's-get-stoned party attitude belonged to a majority of the under-thirty crowd, and that included me.

One night, in the two-bedroom apartment we rented in Toledo, I pulled out the baggie of pot and showed it to Beth.

She wasn't shocked, just surprised. "Where did you get that?"

Beth and I were alike in that way. We would try anything. Out came the Zig-Zags and smoking a little pot now and then became smoking a little pot often.

Only six weeks after starting my new job at WOHO, I got a phone call.

"Is this Ron Copeland?" asked a man with a deep, baritone voice.

"Yes it is."

"This is Paul Cannon. I'm the Program Director of WKNR Radio in Detroit. On New Year's Day, I was sitting at home tuning up and down the radio dial, checking out the competition, when I picked up WOHO. I liked what I heard. At the time, I had no openings, but I do now. Are you interested?"

A verse from the Beatles song, "A Day in the Life," popped into my head. *Somebody spoke, and I went in to a dream.*

WKNR, or as it was affectionately referred to, Keener 13, was a legendary Top 40 radio station, located in the fifth largest city in the nation, and they wanted me.

Only five years out of high school.

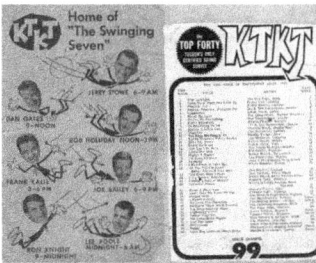

When I was using the name
Ron Knight

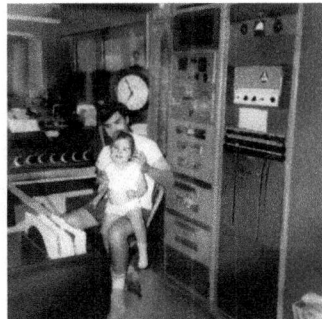

My daughter Rhonda never had
a chance but to spend her entire
adult career in radio!

Clowning around in the studios
of KTKT Tucson, Arizona
in 1967

The Red Dog Saloon in Old Tucson

CHAPTER 6

"PARADISE BY THE DASHBOARD LIGHTS"

Off we went to Dee-troit City. We had been in Toledo for only six weeks, yet we were on to a new adventure. See America. Meet new people.

When we arrived in Detroit, it was less than six months after the infamous riots. The town looked like Britain after World War II. Burned-out buildings were everywhere. It was a mess, but the only place Detroit could go from there was up. After all, they did have Ford, Vernor's Ginger Ale, Tiger baseball, and Motown Records. I had made it to a great American city.

We rented a two-bedroom house about three blocks from WKNR in Dearborn. Right away, we started looking for unusual items to decorate it with and opted for the opium den look. In a used furniture store, we came across a four-foot tall, pure brass Egyptian vase that had a long neck, a wide, bulbous bottom, and a long skinny spout. As far as we knew, it had no function whatsoever, but it was a real conversation piece when we were sitting around stoned with friends, which was pretty often. Beth crafted a giant floor pillow. It was five-feet square and made of a soft, cotton fabric with a zebra pattern on it. Many a friend who was too fucked-up to drive home crashed on that pillow.

I was now a Keener jock, working with fellow DJs Mark Allen, J. Michael Wilson, Dan Henderson, Gerry Goodwin, Ron Sherwood, Tom Neal, Gary Granger, Alan Busch, and Dave Forster. When the day came for my first shift, I was sitting in the ready room a couple of hours before air time, familiarizing myself with one liners and station slogans when Paul Cannon, the program director, came in.

"So, what name do you use on the air?" he asked. "We'd prefer you used an alias."

"Well, I hadn't really thought about it," I said.

Paul reminded me I had two more hours before I went on the air. "You have plenty of time to think one up."

I started running through names I'd heard through the years. Shane Todd, the name of the lead singer of some band I knew back in Ohio was catchy. I came very close to being Shane Todd. A one-syllable first and last name seemed too short to me, so I kept thinking. *Shane Conner, no, Sean Conners. Oh wait, Sean Connery.* I loved James Bond movies, but I couldn't just use his name without altering it somehow. *Wait, I've got it. Sean Conrad. Sean Conrad. That rolls out nicely, and it's as close as I'll ever get to being an international spy for the British Crown.*

At WKNR, the definition of working full-time changed from a forty-hour week to around a thirty-hour week. Monday through Friday, I'd do a four-hour on air shift followed by about an hour's worth of production. Then, on Saturday, a five-hour on air shift. Air talents in a major market were paid a salary versus an hourly wage. This left plenty of time for doing record hops for additional income and to pursue my desire to manage a successful rock band. Watching Bob Holliday manage bands on top of all his other duties as a PD had made a lasting impression on me.

Between 1966 and 1968, I never lost contact with the Jacemen, the band I was managing in Rockford. So about six months after arriving in Detroit, and with all that time on my hands, I figured it was time to get them recorded. I had no idea what I was doing, but that had never stopped me before. Except for my six-week First Class License training, every bit of radio and record industry knowledge I had gained in the past was acquired by just doing it. Jump in with both feet and fake it until you make it. That was my mantra.

Before I brought the band to Detroit for a recording session, I wrote the lyrics to what I thought could be a hit record. It was called "Don't Take It Out On Me." Because of our fast paced lives, there was stress and tension in my relationship with Beth, and arguments became commonplace. We were both screwing around by then. I wrote the song as a way of easing my conscience. I had written the lyrics, and band member Gary Alexander put it to music. It was done as a slow ballad and was pretty damn good, I thought. I was sure that with a little luck it could be a hit single.

The time came to find a recording studio. Because of the success of Motown, Detroit seemed to have a studio on every corner. My, oh my, which one should I use? I opened up the Yellow Pages to recording studios, closed my eyes, and dropped my finger onto one of the pages. It landed on the name Terra-Shirma Studios.

That random decision led me to a meeting with Russ Terrana. He and his twin brother, Ralph, owned Terra-Shirma

Studios. Both of them were original members of a popular De-
troit group called the Sunliners, and Pete Rivera (Peter
Hoorlebeck) was their drummer. Russ played lead guitar, and
Ralph was on the keyboards. The Sunliners later morphed into
Rare Earth.

When I arrived at the studios with the Jacemen, it was apparent I needed help producing the song I had written. Russ
was a master recording engineer and musician. He took me
under his wing and got me through it. I found out later that
right after I got to the studio, Ralph had taken Russ aside and
said, "Russ, this guy is a big time disc jockey on WKNR. Be
cool. Go help him out."

A lifelong bond developed from that session between
Russ and me, and before I knew it, Russ, his wife Joanne,
Beth, and I were joined at the hip every weekend. Getting
stoned and going to concerts, or getting stoned and playing
cards, or getting stoned and getting the munchies, and—well,
getting stoned was what we did.

We saw the live stage play, *Hair*, together—stoned out of
our minds, of course. It was that night at *Hair* that Russ was
bowled over by two cast members, a girl named Stoney and a
guy named Meatloaf. Within a few months, Russ and Ralph
coaxed them into their studio and recorded a great LP. The
song that was to be the single on the album called "What You
See Is What You Get," and it was a real butt-kicker.

Flip Wilson coined that slogan on Laugh In. Unfortunately, a group called the Dramatics put out a song with the same
title, which became the hit, and the Stony and Meatloaf version fell in to the dust heap of records that were really good
but just didn't make it. Meatloaf later became a huge success
on another label with his LP, *Paradise by the Dashboard
Lights*.

Russ's engineering skills for artists like Isaac Hayes, Frigid Pink, Mitch Ryder, George Clinton's Funkadelic, and the infamous MC5 at Terra-Shirma caught the attention of Motown Records. Soon, some of Motown's most successful producers like Norman Whitfield and Hal Davis were borrowing Russ for their sessions. In 1968, he became a fulltime employee of the Motown hit machine. Many times, I'd sit in on his sessions and watch as Russ recorded Smokey Robinson, Rare Earth, the Temptations, and the Jackson 5. A short time later, I'd be playing them on WKNR. I knew this was only the beginning.

1968-9 Keener Bumper Sticker

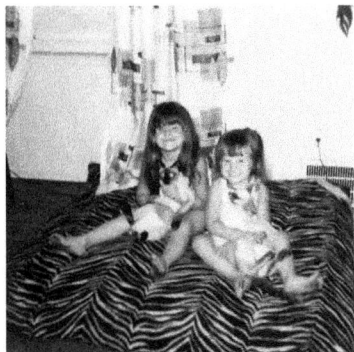

Rhonda, Robyn, and the zebra-skin-patterned crash pad pillow

The Sunliners who morphed into Rare Earth–Left to right: Ralph Terrana, John Persh, Russ Terrana, Gil Bridges, Pete Rivera, and Steve Fisher

On the air at WKNR.
Check out the antique
equipment!

My buddy Russ working his music
magic at a Motown mixing
board in Detroit

CHAPTER 7

"KICK OUT THE JAMS"

As I had done everywhere else I worked, I started a record hop business right away and spent most weekends spinning discs for the many high schools in the area. I had two huge speakers, custom built for me by Russ. I also purchased spotlights that sat on tripods and double turntables to allow me to create the effect of a live radio show. In order to cart all this equipment around, I bought a 1959 Pontiac Hearse. What a boat that was! I loved that car and drove it to work every day. We would take the girls to the drive-in movies in it now and then.

One time, we pulled in to a space next to a car with an older couple in it. The movie was a slasher horror flick, and a few minutes later, the couple looked over at this hearse, started their engine, and moved to another space farther back. Between the cab and the cavernous open space where the casket would sit was a sliding glass window. When we slid the window shut, the girls could laugh and play, while we enjoyed the movie in relative peace and quiet.

Periodically, when a new up-and-coming artist was being promoted by a record company, the company would arrange for the new artist to appear at one of my record hops and lip sync the new record. When Neil Diamond was promoting "Cherry Cherry," he stopped by one of my hops. Because we knew in advance that he was going to be there, I hyped it all week long on the air. We packed the house with high school kids.

Neil walked on to the stage with guitar in hand. I introduced him, he did his thing, and ten minutes later, he was out of there. The kids loved it, and I made two hundred-fifty dollars that night, which was a lot of money in the late 'sixties. Poor Neil made nothing. I guess you could say he did okay in the long run, though.

We became regulars at Detroit's famous Grande Ballroom during this era. The Grande (pronounced gran-dee) was built in 1928 and boasted an outstanding hardwood dance floor. During the '30s, '40s and '50s, it was the place to go for big band music and showcased artists such as Glen Miller and Harry James. In 1966 it was acquired by high school teacher and local DJ Russ Gibb and became a venue for Detroit style rock and roll. The Grande's regular, local, house bands included the Stooges, James Gang, Bob Seger, Frost, SRC (Scott Richards Case), and the MC Five. Eventually big name national groups also started appearing at the Grande.

The MC 5 was the loudest, rawest, head-banging rock and roll band I had ever heard. When they played at the Grande, which was often, the sound was deafeningly loud with ample distortion coming from the guitars and vocals. They were managed by John Sinclair, founder of the White Panther Party, a militant leftist organization *of white people working to assist the Black Panthers* (in their own words). The MC 5's radically charged, over-the-top brand of music gained them national notoriety.

They were embraced by the young and naive Grande Ballroom audience who, of course, were blotto stoned on acid and pot. To us stoners, the walls appeared to flex and pulsate in a constant state of motion to the sound of the music. This liquid light show, consisting of colored mineral oil and alcohol, washed over the entire ballroom with what looked like a tie-dyed T-shirt in constant motion. The pungent smell of patchouli oil mixed with reefer smoke clouded the air. We bought into Sinclair's radical movement hook, line, and sinker, endorsing it but never getting involved. It was cool, man, know what I mean? We even went to a White Panther Party celebration at Sinclair's crash pad for the release of a new MC 5 LP.

They were also notorious for the lyrics to the title track of their LP, *Kick Out The Jams,* which was recorded live at the Grande. I was in the audience when it was recorded. The opening salvo of the song is the lead singer screaming out, at the top of his lungs, "Kick Out The Jams, Motherfucker!"

The executives at their label, Elektra Records, were painfully aware that if they expected to get it played on the radio, the *motherfucker* in the lyrics had to go. Rob Tyner, the lead singer, was sent in to Russ and Ralph Terrana's Terra-Shirma Studios to re-record the opening as "Kick out the jams, brothers and sisters." Russ razor-bladed the new line in to the existing song. It worked because even though it was never anything close to a mass-appeal hit, it did get Top 40 air play. We played it at Keener.

The group lost its contract with Elektra when the conservative, one-hundred-year-old Detroit department store, Hudson's, refused to carry the album, and the band took out an ad in the radical newspaper, Fifth Estate. The ad read simply, "Fuck Hudson's." The Elektra logo was prominently displayed in the ad, and Hudson's pulled all Elektra products off the shelves.

Walking into the Grande on a Saturday night, stoned to the gills, was a reflection of the times. Located at the corner of Grande River Avenue and Joy Street, the Grande's spacious hardwood dance floor was where the action was every weekend from 1966 to 1972. The tie-dyed freaks would come out to spin and twirl to the sounds of the Grateful Dead, Janis Joplin, The Who, Jeff Beck, Led Zeppelin, and Cream. The Grande was Detroit's answer to San Francisco's Fillmore West.

One night, wearing my new Dennis Hopper Easy Rider rawhide buckskin jacket, and escorting Beth in her long flowing paisley dress, I attended the first American performance of The Who's *Tommy Rock Opera*. We were absolutely blown away by the unknown artist who opened the show. Joe Cocker came on and mesmerized the audience to the extent that, by the time The Who came on, it was anticlimactic. In 1972, the Grande closed. It's rarely been used since and has fallen into a state of extreme disrepair.

In 1968, we had tickets in the first row, dead center, to a Jimi Hendrix concert at Cobo Arena. Jimi was no more than ten feet away from us. When the show ended, Mitch Mitchell, the drummer, tossed a drumstick straight in to the crowd. It was headed toward Beth's head when I brought my hand up and caught it just before impact. Even stoned, my reaction time was still excellent at that age, and it was a thick, heavy drumstick. It had to be, to be heard over Jimi. We did get to witness the burning of that guitar. At the end of the show, Jimi's final encore song was "Purple Haze," it brought down the house.

At another concert at Cobo, all of the Keener DJs were standing on the stage getting introduced to the sell-out crowd, as Jim Morrison walked past us to front a Doors concert.

The squeaky clean image I projected on the air was the complete opposite of what the Grande Ballroom was all about. Psychedelic, underground FM radio was starting to pick up speed. WKNR-FM, one of the first underground stations in Detroit, was situated about twenty steps down the hall from WKNR-AM, and the two stations were programmed to appeal to completely different listeners. The pay was much better on AM radio, but my heart was on the FM side. Since we were all part of the same company, I would fill in on FM regularly. Imagine getting off the air after four hours of high energy, top 40 jocking, then walking down the hall and doing another four-hour shift, but with a delivery that was intended to make me sound stoned. Most often, I really was stoned. One time my regular AM shift overlapped with an FM fill-in shift by one hour. I can attest to the fact that running up and down the hall from one studio to the other, switching personas for one full hour is mentally exhausting.

Beatle fans have heard about the *Paul McCartney is dead* rumor that surfaced in 1969. I was hanging around after my AM shift one evening with Russ Gibb, who was on the air on WKNR-FM, and I witnessed this whole thing unfold right before my eyes. A caller said he was sure Paul McCartney was dead. He asked Russ to manually play "Number 9" backwards, and when he did, it sounded like "Turn me on dead man."

That one call opened the floodgates to call after call from people with yet another clue as to why they were sure McCartney was dead. One caller pointed out the fact that on the cover of the *Abbey Road* LP, Paul was walking out of step with his band mates and was barefoot. Another caller mentioned that if you strained to listen, you could hear John saying "I buried Paul," at the very end of "I Am the Walrus."

Soon the national news media got hold of the story, and it exploded across the planet like a rocket to the moon. We even got a long distance phone call from someone who claimed to be Paul, assuring everyone that he was indeed alive. We never knew if it was actually him, but the voice was right. It was exciting to be there at that moment in time and see Beatle conspiracy theorists whipped into a frenzy.

No words could correctly explain the joy of sitting down and listening to a new Beatles LP from beginning to end for the very first time. Because I was in the radio business, I was always in a position to hear a new product before the listeners. It made me feel honored to be the first to turn the audience on to, what we felt at the time to be important, message-laden music. At this time, in the career of the Beatles, almost every song they wrote had a message you could read into it if you wanted to. Charles Manson heard there was going to be a race war in "Helter Skelter."

One of my emcee assignments in 1968 was to be the announcer for a station promotion at Hudson's Department Store one Saturday afternoon. Appearing live was the Amboy Dukes. From that encounter, I became acquainted with Ted Nugent, who at the age of only eighteen, was the Amboy Dukes' lead guitarist and still living at home with his parents. Ted was mature beyond his age and never dabbled in drugs or alcohol. He was an accomplished self-promoter and appeared regularly as a solo artist all around the Detroit area. I drank a Coke in his kitchen while his mom was busy preparing dinner. His showmanship was over the top, like the time he swung onto the stage, Tarzan style, on a rope from the balcony, swinging himself only a few feet above the audience.

He was wearing a tall, white hat with long ties hanging down and fuzzy white pom-poms on the end of each tie. And nothing else but a loin cloth. He landed on the stage, grabbed his guitar, and slammed in to a guitar solo as only Ted could do.

He was an incredible businessman, even at that young age. He once said no matter how much he made in a given week, he would take just fifty dollars in pocket money and bank the rest. That kind of level-headed thinking was uncharacteristic of rock stars in the 'sixties. Not to mention disc jockeys.

The AM and FM studios were just a few steps away from each other. One afternoon Paul Cannon walked in to the studio during my shift.

"Grace Slick is down the hall being interviewed on the FM," he said. "Would you like to meet her?"

Besides the fact the Jefferson Airplane represented what was cutting edge cool at the time, Grace was a stone fox. Did I want to meet her?

"Uh, yeah!"

I wanted so badly to be accepted as cool by her, even though our format did not allow for a momentum-killing pause for something as low key as an interview. The underground groups of the day depended on us Top 40 stations to cross their records over to the masses, but we represented the establishment to them. We were way un-cool. So with Grace Slick standing behind me in the studio and with great apprehension, I turned on the microphone and started bellowing out my introduction to the next record.

"Keener 13, with another Cash Call Giveaway coming up in the next hour! Stay tuned and listen for the…blah, blah, blah…"

As I was doing this, Grace started yelling in the background at the top of her lungs, "Shoot that man some more speed! Shoot that man some more speed!"

It was humiliating, but I got to meet the queen of cool, Grace Slick.

The power to make or break a new record was in the hands of the PD, not the on- air talent. But the artists didn't know that. An example of this was an evening in 1968 at the Moon Supper Club in Windsor, Ontario, just a short drive across the Detroit River in Canada. Kenny Rodgers and the First Edition were appearing live and, as usual, we had free tickets to see them. After the show, the record company promoter took us, along with Big Jim Edwards and his wife Sylvia, backstage to meet Kenny and the group. "Ruby, Don't Take Your Love to Town" and "I Just Dropped In To See What Condition My Condition Was In" were huge hits at the time. Kenny was as nice as could be and made us feel very welcome.

"Sean, I want to thank you for playing our records," he said. He started telling me about their next release and how he hoped we'd play that one, too.

A few months later, just days before Christmas 1968, I was at home with the family when the phone rang. Beth answered it and said, "You won't believe this. It's Kenny Rodgers, and he wants to talk to you."

"Huh?" I had no idea why he would be calling me or, for that matter, how he got my phone number.

"Hi, Sean, this is Kenny Rodgers. I'm in Hollywood shooting a TV special for ABC, and I had a break in the action for a few minutes and thought I'd give you a call to wish you and your family a Merry Christmas."

Apparently, he thought I could play his record whenever I wanted to, when in reality, I had no control over it ever being added to the playlist. I didn't tell him that, though.

The Grande Ballroom as it appears today.

Inside the Grande Ballroom today

Christmas 1968

Our opium-den-look living room in Dearborn with the Egyptian vase added for effect

My beloved 1959 Pontiac hearse used to transport record hop equipment

A Keener music survey from 1968 with my good friend Mark Allen (Bob Dearborn) pictured

CHAPTER 8

"SPILL THE WINE"

From the day I was hired at WKNR, it was obvious to everyone that the station was for sale. It wasn't talked about openly but was an underlying rumor around the building. This affected employee morale and limited the money available to operate the station. The promotion budget was minuscule. We were in direct competition with CKLW, a Canadian radio station just across the river. They had no budget restrictions, and that made it difficult for us to go after them in the ratings.

Another disadvantage was signal strength. We were a 5,000-watt station licensed to Dearborn, Michigan that went low power after sunset while CKLW, not ruled by the FCC, was at 50,000 watts twenty-four hours a day.

Probably the biggest disadvantage we had was the most important one, programming. CKLW was programmed by the hottest team in America, the Drake-Chenault organization.

Programming wiz-kid Bill Drake and Gene Chenault, who handled the business end of the company, had contracted with the RKO General Radio Corporation to have absolute, total control of the programming on all their radio stations. The Boss Radio format was heard on KHJ, Los Angeles; WOR-FM, New York City; WRKO, Boston; WHBQ, Memphis; KFRC, San Francisco; KGB, San Diego, and KYNO, Fresno. Boss Radio was conceived at KYNO, which was owned by Chenault. After two years of beating my head against an immovable object, and tired of the imaginary for-sale sign I saw every day when I came to work, I decided that the time had come to work for Drake-Chenault.

In 1968, Beth and I were seated next to Big Jim Edwards and his wife Sylvia at a dinner party. Big Jim was the noon to 3 p.m. jock on CKLW, and he was big! At least six-foot-seven, he was a gentle giant with a booming voice. When he cut loose with his famous laugh, the walls would shake. We became great friends, and soon the four of us became the six of us with Russ and Joanne from Terra-Shirma. We had two kids each, and we would arrange for one babysitter to watch all six of them. That freed the six of us to attend three of the most popular showings of the late '60s: The stage play, *Hair*, the opening night of the Beatles *Yellow Submarine*, and Woody Allen's *Take the Money & Run*. We were full of piss, vinegar, and marijuana, and we were having a ball.

Jim and I were competitors, but that never affected our friendship. One day I called to pick his brain about something that had been on my mind for a while.

I started off with. "Hi, Jim, what's hapnin?"

"You, baby, you're happening."

"Yeah, right. Hey, I got a question for you. I'd like to make contact with the Drake-Chenault organization and was wondering what you thought would be the best way to do that."

"Why?" he asked.

"I'm really tired of the bullshit over here, and I think I'm ready to move on to another station. I'd love to work for Drake. Who do I talk to?"

"Well, probably the best thing for you to do is contact my PD, Paul Drew, and tell him about it," he said. "I'm sure Paul will point you in the right direction. Besides, I know he'd love to screw with WKNR by getting you out of the market."

Paul Drew was well known for being difficult. He always had a small earphone in one ear and never missed listening to the station. When he slept, he would tape record every moment so he could later go back and speed-listen to what he missed while he was sleeping. Make the slightest mistake on the air, and the bright red bat phone sitting next to the jock would light up immediately.

He'd bark in to the phone, "How come you missed the logo? Are you tired or something? Wake up! No mistakes."

So, calling him was not something I looked forward to. Nevertheless...

"Hello, Mr. Drew," I began. "This is Sean Conrad. I'm the afternoon drive jock at—"

"Sean Conrad, I know who you are. So, why are you calling me? You tired of us kicking your asses over there?"

"Well, I guess you could say that," I replied. "All I know is I'd love to work for the Drake chain. Anywhere."

"Let me put it this way, it won't be CKLW. We've got a solid staff line up, but I heard there's an opening at the flagship station, KYNO in Fresno. Would you consider there?"

You mean, California, I thought? *Hell yes! I'd take a job anywhere in California.* "Well, sure."

"Okay, send a tape and resume to Harry Miller, the PD at KYNO. Tell him I told you to contact him."

"Okay, Mr. Drew, thank you."

"Don't thank me," he said. "You haven't got the job yet. Let me know what happens."

Off went the tape and resume, and in March of 1970, we left for California. I loaded up our gold 1969 Opel Cadet station wagon with two adults, two kids, and two big Siamese cats. The cats had a litter box in the back part of the car. We had a handful of joints and a bag of *Christmas trees*, which was slang for pharmaceutical speed. Just before we left, I convinced my doctor to write me a prescription for a supply of amphetamines so I could lose weight. It's amazing how a dangerous drug like that was so easy to get back then. There we were, in our middle twenties, headed west, smoking pot, and gobbling speed all the way to Fresno.

KYNO was located on North Barton Avenue, less than a half-mile from Cedar Lanes Bowling Alley. In Fresno, in the late '60s and early '70s, Cedar Lanes was the best place to be in the evening if you wanted to eat really fine cuisine and get blotto at the same time. Many radio station promotions and game plans were drawn out on napkins at the Cedar Lanes bar by Bill Drake and his staff.

That bowling alley had the world's best prawns and prime rib. The Cedar Lanes coffee shop was open all night. It was where we went for breakfast when we were ready to come down from the evening's debauchery. The 1950s-style cocktail lounge had embossed, red velveteen wallpaper, and black vinyl covered half-circle bench seats.

No less than three months after our arrival at KYNO, Harry Miller took a job at KGB in San Diego. I thought, what the hell, I'm going to toss my hat into the ring to be his replacement. As far as I was concerned, that had been my life's calling all along. I wanted to be a PD just like my mentor Bob Holliday. I hadn't been there very long, so I tried out my idea on the chief engineer Dave Evans. Unlike most radio station engineers, he was one of us. An audiophile beyond belief, Dave was our go-to guy when it came to buying the best home stereo equipment. He was also eccentric. Very few people back then drove Citroens, the funny-looking French car that looked like an upside down bath tub. Dave had one, and I bought one myself.

Dave could out-party any of us but was also known company-wide as a genius when it came to fine tuning an AM radio station to sound better than any other in the market. After smoking a joint one night, I was pounding down a few with him at Cedar Lanes.

That night, with that unmistakable cigarette-and-floor-polish-bowling-alley smell in the air, I told Dave I wanted the job of PD.

"I think you should go for it, but get on it right away," he urged me.

"I was going to talk to Decker about it tomorrow," I said. "He's the general manager, after all."

"Forget about Decker. Go straight to Chenault. Believe me, Decker has no say so in the matter, and if you can sell Chenault on the idea, he'll highly recommend you to the Drake people in LA."

"It's kinda scary calling the owner like that," I replied. "I've only been here three months."

"Do it," Dave said. "Gene's a great guy, and he'll be impressed if you go straight to him."

Gene Chenault was not only a savvy businessman, he was, as Dave had told me, a great guy. When Gene was in town, he'd go from desk to desk and office to office just to say hi and spread goodwill. He always had a smile on his face and made his employees feel appreciated. Bill Drake would have never reached the level of success he had without Gene out front, running interference for him. Gene made damn sure the sales effort came second to the programming effort, and that was a crucial ingredient of the Drake format. Before Drake, sales ruled, and there were few limits to commercial content in any given hour. Drake restricted how many commercials could air, which created more time for music and jock input.

"Hi, Mr. Chenault," I said when I finally got up the nerve to call him. "This is Sean Conrad, your noon-to-three guy here in Fresno. Have you got a second? I have something I'd like to talk to you about."

"Well hello, Sean," Gene replied. "How's the weather up there today?"

"Well, it's supposed to hit 105, but I love the heat. We didn't have a lot of that in Ohio."

"So, what can I do for you, Sean?"

I blurted out my reason for calling, nervously babbling along, and made my case as to why I should be his next PD in Fresno.

"Well, you sound to me like you could do the job," he said. "All I can do is suggest to the Bills that they give you an interview, because they are in charge of that decision. I will tell them about you, though. Phone me back in a few days, and please, call me Gene. Mr. Chenault was my father."

The Bills. Bill Drake and Bill Watson. If I were to secure the interview, I knew that getting past Drake would be the easy part. The absolute scariest part of what I was about to go through was the interview with Watson, which was the only name anyone ever called Bill Watson. He was the national PD for the chain, and was Drake's right-hand man, handling all the gory details of programming eight radio stations scattered across the country. He was intimidating and could be a real hardass. He wore open-collar silk shirts, always with some sort of gold jewelry around his neck, and tinted prescription glasses he'd peer down at you through.

I called Chenault back the following Monday.

"I have you booked on a flight to LA tomorrow morning at eight," he said. "Harry has someone to fill in for you, and we'll send a limo to the airport to pick you up."

"Tomorrow morning, that's great. Thanks, Gene, I'll see you then."

As I was being escorted in to Watson's office the next day, I hoped no one could hear my knees a-knockin' or see the nervous twitch going on in my left cheek.

I introduced myself, and Watson unceremoniously said, "Sit down."

I took a seat in front of his desk, feeling small. It was as if his desk was perched on a six-inch high riser. The soft leather sofa directly opposite that monstrous desk swallowed me up, so that my knees were almost level with my eyes. I felt a trickle of sweat just above my right eyebrow.

"So, what makes you think you can handle the job of PD?" Bill asked as he exhaled a puff of smoke from his Benson & Hedges.

"For one thing, I know what it takes to motivate people," I squeaked out.

Silence. A good sixty seconds of deafening silence. In that situation, sixty seconds felt like sixty minutes.

"So how do you know that?" Bill finally said.

"Well, I know how I like to be treated by a program director and talking down to me or making me feel like I could easily be replaced is not the way to do it."

More gut-wrenching silence. This went on for at least a half-hour. When I left that room, I was drained.

A week later, Wayne Decker came in to the studio while I was on the air and asked me to stop by his office when I got off.

Uh-oh I thought. *Am I in some kind of trouble? Or, maybe it might be...Damn, another hour before I'm off the air. This is driving me crazy.*

I greeted Wayne at the door to his office.

"You wanted to see me?" I asked timidly.

"Yeah-yeah-yeah, sure, sit down. You got the job. You start Monday. Miller's last day is Friday."

"That's great! Should I tell the staff, or will you?"

"I scheduled an all-staff meeting Monday morning," he replied. "I'll make the announcement then."

I'd been on the staff only three months, and I knew my fellow jocks were taken by surprise when the announcement was made. The rumor at KYNO was that someone from out-of-the-market would be their new PD. It took several months for me to gain their confidence. When they realized I was the kind of PD who treated them with respect and always looked out for them, they got past any apprehension they had about my rapid promotion.

I frequently scheduled weekly jock staff meetings to take place over dinner and drinks at Cedar Lanes. Food fights were commonplace at these staff meetings, and more than once, we came close to being asked to leave because of our rowdiness.

My air staff at one time or another consisted of Dirk Robinson (Dirk Raaphorst), Joe Angel, Steve Randall, Bill Stevens, Pete McNeal, Chip Roberts, Ted (Kraft) Jordan, Todd Walker (Greg Crawford), Mark Daniels (Dave Yodelman), Chris Van Kamp, and Mike Novak. Some of these guys had to go through the Sean Conrad "welcome to KYNO" ceremony at Cedar Lanes. This required having to pound down shots of the deadly 110 proof, Green Chartreuse liqueur, considered by some to be the nastiest, bar drink ever. It was the Jagermeister of its time. Tasting like furniture polish mixed with sugar, it was made from herbs and kicked like a mule.

None of us were older than twenty-five, and we still had a lot of kid in us. Innocence is really what it was. Because we commanded huge ratings that fueled that sales effort, we were insulated from most of the radio politics and were, in ways, quite naive.

A quick stop in Rockford to visit
Dad then off to California

My good friend Harry Miller
who hired me at KYNO in
March of 1970

My upside-down bathtub Citroen

Cedar Lanes

One of my favorite people,
Steve Randal

KYNO jock initiation liquid

CHAPTER 9

"ONE TOKE OVER THE LINE"

Fresno from the mid-sixties to the mid-seventies was a major test market, especially for music. When Bill Drake began programming his version of what Top 40 radio should sound like in order to garner the largest audience, it was a radical concept. In a short time, KYNO was the radio station of choice for as much as 40 to 50 percent of the available audience, leaving most other stations in the dust. Since a massive listening audience translated into titanic advertising dollars, the next step was to take the Drake format to Los Angeles, where it repeated its success within a few months of launch.

In no time, Bill Drake was programming the entire RKO General Radio station chain. Soon, every city in America had a Drake knock-off, sound-alike format. This gave him the power to make or break a new recording artist. Many times, he would first test a new record out in Fresno. If it made it there, it could make it anywhere.

It was exciting to add a new, untested record by an unknown artist at KYNO. When we did, we had to make sure the record was in the stores. If it wasn't, there would be no way to judge its appeal to the radio audience. We had to see sales activity in order to push a record up the weekly KYNO Top Thirty charts. Those chart levels would then be reported weekly to all the national trade magazines like Billboard, Record World, and the Gavin Report.

Radio stations across the country would watch those charts carefully and base their weekly ads on what they saw. When I first heard "Your Song" by Elton John in 1970, I was blown away. KYNO was one of the first stations to add it, and as it started climbing our charts, stations everywhere began playing it. I'm sure our assistance in creating a fan base in Central California early on in Elton John's career was why Fresno was one of only four cities in California included in his Spring 1971 Tour. He played Fresno May 12, 1971 followed by San Francisco, Sacramento, and Anaheim. My brother Jim was in town and attended the concert with us. Fresno loved Elton John and exploded into applause when he played "Amorena," an obscure album cut we had in heavy rotation on KYNO. Even though KYNO was in an agricultural town with triple-digit heat in the summer, up and coming program directors and disc jockeys would do anything to go there. This also made KYNO an electrifying place to work.

As program director of KYNO Radio, I partied so much that I'm lucky to be alive today. The record companies sent their emissaries from the big city to Fresno regularly to nudge us into adding their records to the playlist. If you call drugs, hookers, eating at the finest restaurants, and the star treatment backstage at concerts in Los Angeles and San Francisco payola, I guess I can't argue with you. But, to me, payola was receiving cold, hard cash, which never found its way in to my pockets.

One night, Kenny Reuther, a promo man from San Francisco came to town to work his record, and work it he did. Four of my jocks and I were wined and dined at an expensive Fresno eatery. By midnight, we were all blotto, and after dinner, someone suggested we go to the Belmont Massage Parlor for fun and games. When we staggered up to the front door, a guy opened it a crack and said they were closed. Well, money talks and bullshit walks. Kenny cut a deal with the guy. We all walked in, grabbed a girl, and all of us radio personalities got to experience something one only sees in a triple X movie. I don't recall if we added his record or not.

Three or four times a year, the energy level shot straight up because those were the times Donny Branker would come to town with his rock and roll circus. Don Branker was a West Coast concert promoter, who later went on to produce California Jam's One and Two in 1974 and 1978. Cal Jam One drew 400,000, and became second only to Woodstock in concert crowd history. Hundreds of thousands of gyrating, stoned-out freaks were rocking out to the music of Rare Earth; Earth, Wind & Fire; The Eagles; Black Oak Arkansas; Seals and Crofts; Deep Purple; and Emerson, Lake & Palmer.

Even though Don traveled the country, wheeling and dealing rock and roll, he always made time to come back to Fresno to promote concerts since it was his home town.

It went like this: I'd get a call from Don...

"Sean, I'm bringing Redbone and Flash Cadillac to Selland Arena on the twenty-ninth, and we need to start up the promotion machine now. I'll see you Monday."

Or, "I've got Tower of Power and Elvin Bishop coming to the Rainbow Ballroom in August. Put it on your calendar."

Don would come charging into my office, and we'd start up the hype machine with on-air ticket giveaways, high-energy promotional announcements, and any other gimmick we could come up with to give him as much free promotion as possible. Those free ticket giveaways worked to the benefit of the station ratings, too, so we did whatever we could for Don. Around 1972, an FM underground station hit the airwaves, and we did not want them to take his advertising dollars. In some cases, he had to go with them, however, because some of the hard rock groups he brought to town were not on the KYNO playlist.

Don and I would spend hours in the KYNO production room creating sixty-second commercials that would whip people in to a frenzy to go get their tickets before the show sold out. In most cases, they did. We'd fire up a joint and let the creative juices flow.

That was the business part of Don's regular visits. The pre and post parties at the Smugglers Inn and, especially, the after-concert all-night parties were legendary. All the groupies and hangers-on would be there with ludes, reds (slang for Seconal), or downers as they were called, pot, and lots of booze. Since Pat and Lolly Vegas and Redbone hailed from the Fresno valley, they always had the largest gallery of groupies. Periodically, the KYNO jocks would play baseball against Redbone and their roadies.

As Don always said, "You may hit a home run, but good luck getting past second base." You'd get knocked on your ass long before you could round third base.

In 1972, I saw a great opportunity to promote the station and register as many new young voters as possible for the upcoming election. Most young people hated Nixon, who was running against McGovern, and this was our way of trying to save the world from Tricky Dicky. If we could coax a young kid to register to vote at the concert, we assumed the party they'd register with would be the Democrats. One of my closest Fresno friends, Conrad Jimenez, hooked up with Bob Kostave and Russ Bader, two insanely radical Fresno State students, who wanted to use this event as their final thesis to graduate.

With a lot of help from Don, we put on a free rock concert at Radcliff Stadium. It drew 25,000 people and made the national news wires. With Don's power in the record industry, he convinced War, The Doors, Sweathog, and Dr. Hook to appear free at this concert. There was no admission charge, and voter registration booths were strategically located throughout the stadium.

KYNO benefited greatly from it in the next ratings book. Nixon won anyway.

Around the summer of 1971, I was buzzed from the lobby by the receptionist.

"Sean, there's a man here who wants to apply for a job. He has a tape and resume."

"Well, I'm in the middle of something right now, but, oh well, I need a weekend guy. Send him in."

He introduced himself as Steve Cary. I asked him to have a seat and hoped I sounded friendlier than Bill Watson had sounded to me.

After listening to his air check from KMEN in San Bernardino, I told him he sounded pretty damn good, and that I had a Saturday and Sunday night shift available.

"I'd love it. I'll take it. When do I start?"

After Steve left, I decided to change his on-air name to one that would sound better on the air. It's one of the perks of being a program director.

You may wonder how disc jockey aliases get dreamed up. In Steve's case, one of our salesmen at KYNO, a great and beloved old timer, Don Randall, walked by as Steve was leaving.

Perfect, I thought. I had found his new on-air name.

When he reported for his first shift, he looked at the weekend line up sheet and saw the name Steve Randall. Confused, he tracked me down and asked, "Who is this Steve Randall guy?" I laughed and told him, "That be you, Bubba." That has been his legal name ever since.

That Saturday night, Steve Randall hit the airwaves. A funny guy with a great voice and bizarre sense of humor, he could do Gregory Peck better than Gregory Peck. His Wolfman Jack was so flawless that I would swear he was da woofman. Within six months, I had Steve doing one of the most important air shifts of the day, afternoon drive, 3 to 7 p.m. One evening, he was on the air and he answered the request line in his Wolfman voice. He did that a lot.

"Hello, who's dis on the Wolfman telephone?"

"Wolfman? What are you doing there? You're in Fresno?" chortled the hitline chick.

"I was just passing through and thought I'd stop by and say hi to my friends at KYNO. What's your name, and what can I have Steve play for you?"

"My name is Karen, and can you have Steve play "One Bad Apple" by the Osmonds please?"

"Sure baby. By the way, how's your peaches?"

"Oh my peaches are great, Wolfman."

"Tell me, uh, Karen, what are you doing right now? Are ya naked?"

"I'm at work at the Kaleidoscope Massage Parlor on Blackstone Avenue." Obviously, there was more than one massage parlor in Fresno.

"Hey Karen, you got any free samples for da Wolfman?"

"Well sure, Wolfman. Why don't you stop by later to-night?" We could hear other girls giggling in the background.

"I will baby, and I'm gonna bring my buddy, Sean Conrad, with me," he said. "Will you show us your peaches? How 'bout we bring you some record albums?"

These girls really believed they were talking to Wolfman Jack.

Steve, sounding but not looking at all like Wolfman and I, with an armful of record albums and hormones a-raging, decided to take a chance and go meet these girls. We pulled up to the curb across the street, where we could see the front door, but they couldn't see us. It was a warm summer night, and every now and then, one of them, scantily clad, would open the door to see if we were there yet.

"Shit, man, we can't get away with this," I said. "They'll know immediately you're not Wolfman Jack. You just don't look like him. Any fool can see that."

"Aw c'mon. We'll just tell them the truth. Let's go."

"No way, let's go get a drink."

That was the end of that. We never got to meet the employees of Kaleidoscope Massage Parlor, but we did get to hang out with some talented individuals.

Encounters with rock stars took on a whole new meaning one day in 1972 when Carl Wilson of the Beach Boys dropped by KYNO. That year, he and his brothers had started their own record label, Brother Records, and Carl was out on a promotional tour with a South African group called Flame. He had signed the group to the Brothers label and wanted to make them stars. And he knew enough about the record business to understand that one of the best ways to get a record played by the entire Drake chain was to get it on KYNO first.

After he introduced me to the band members, had me listen to the Flame's single, and gave me all the reasons why I should add the record to our playlist, the conversation took on a more personal tone.

"Hey, Sean, do you get high?" he asked.

"Oh yeah."

"Well, we have some incredible stuff the guys brought with them from South Africa. You gotta try this shit."

"Sure," I said. "Let's get out of here."

There was no way we could all fit into my old beat-to-shit Volkswagen, so we piled in to Carl's car. This happened to be one of the many times Beth and I were living separately, and since the money was spread thin, I was living in a dumpy apartment in a bad neighborhood, only a few yards from some railroad tracks. None of that bothered Carl and the boys. We had a good time getting stoned and talking trash.

Anyone who knows the history of the Beach Boys is also aware that in the early 'seventies they lived in a mansion in LA that had formerly been owned by Edgar Rice Burroughs, the author of Tarzan. In 1919, Burroughs had purchased a large ranch in the San Fernando Valley, built this home, and later developed the land in to the suburb of Tarzana.

"Sean, next time you come down to LA, stay at our place in Tarzana," Wilson said. "We've got plenty of room, and I know you'd have a great time."

I just nodded. *Carl Wilson*, I kept thinking. *I'm getting high with Carl Wilson in Fresno, California.*

That was the power of KYNO.

Back row: Redbone, including Pat & Lolly Vegas and Don Branker. Front row: Kristi Rohlfing, music director, Randy Brown, Epic Records, and me

Ted Jordan, Elton John, and Sean Conrad

A last-minute power pole planting with Dirk Robinson and Marty Sherwood doing the grunt work

Russ Bader, Bob Kostave, and Don Branker backstage at Vote America 72

From the stage of the Vote America 72 concert

My daughter Rhonda pitching in
at Vote America 72

KYNO 25th Aniversary Patch

CHAPTER 10

"BANG A GONG"

In spite of being a workaholic PD in Fresno, my personal life was a shambles. Beth and I floated back and forth between being together and being separated. We loved each other, but immaturity and the temptations around us in the early 'seventies were too strong to resist. During those times of separation, I did my thing, and she did hers. Fresno was much smaller then, so occasionally our separate lives would cross paths.

Late one Saturday night I found myself at the Fresno Hilton, which had a popular bar and restaurant. I walked in with one of my jocks, Joe Angel, who was also separated from his wife. Beth and Joe's wife were sharing a house together. Joe and I knew they were working as cocktail waitresses at this bar and decided to stop by. We were both a little shocked to see that they were dressed like playboy bunnies and looking hot.

He and I shared a twinge of jealousy that kind of kicked our asses. We got out of there in a hurry.

Because of the expense of living separately, neither Beth nor I had any money. We filed for bankruptcy, wrecking our credit rating. With the car re-possessed, I scraped up a couple of hundred dollars and bought that beat-up, old 1965 VW for transportation. One day, Gene Chenault drove in to the KYNO parking lot and saw me trying to get the car started. Imagine Gene seeing his PD sitting at the wheel of a trashed old bug with four jocks behind, pushing it in order to get it going fast enough to start when I popped the clutch.

A few days later, Gene called me in to his office and said, "Sean, I saw you trying to get that car started the other day. Why are you driving that old beater?"

"Because I'm broke. I have no money," I replied.

"Why? Am I not paying you enough?"

He was very appreciative of the fact that, in spite of my personal life, I was his guy. Programming the station often required me to work into the wee hours of the morning. Sometimes I put together on-air giveaways just before a rating window opened. Other times, I rebuilt the oldies library. Whatever it took to keep KYNO a solid number one in the market, I'd do. And, he knew that.

"Look, Sean," he said. "I'd rather my PD wasn't driving around in a wreck like that. I want you to go out to J.H. Sanders Ford and pick out a car. We have a trade there, so pick out any car on the lot."

"You're kidding."

"No, I'm not. Your last ratings were gigantic, and this is my way of thanking you." It was his way to give me a raise without giving me a raise. Trade at a radio station was a wonderful thing back then.

"But I can't afford gasoline or insurance," I said.

"We have a gasoline trade you can use, and the car will be insured on the company policy. Just go pick it out."

"Thanks Gene," I told him, still in shock. "You don't know how much I appreciate this. What can I say?"

"Just tell me you'll run that bug off a cliff somewhere."

"Done."

In those days, gas mileage wasn't an issue, especially, if your gasoline was free. I picked out a black 1971 Lincoln Continental Mark 4 with a black interior and vinyl roof. This car looked like something a mafia don would drive. It only had 15,000 miles on it, and it drove like a dream. On Monday, I was driving a piece-of-shit rattle trap, and the next day, I was cruising around on a magic carpet.

Scott Cohoe was my music director. I hired him away from KFIG-FM, the only underground station in town. He was a musicologist who stayed on top of new up-and-coming alternative rock artists. Across the nation, free-form FM radio was starting to put a dent in the ratings of Drake's stations so, I was given permission to tamper with the format and test out a few radical programming ideas, such as turning it into an underground station after 7 p.m.

Staying true to form, KYNO once again became the testing ground for format concepts. Drake, Watson, and Bernie Torres, the national music director, were as far away from the hip, new drug music scene as anyone could be. These guys were drinkers. As far as I knew, no drugs. In another time, another place, they would have made great Sinatra Rat Pack candidates.

So we began an experiment, fusing the Top 40 format with the Album Oriented Rock or, AOR format as it became known. I gladly led them into the dark shadows of the rapidly increasing underground music scene. Scott was the epitome of that scene. He was frequently loaded on downers—reds, ludes,

Seconal. Once he was at the station, I started doing them, too, but only after work. These drugs made us slur our words like Tommy Chong of Cheech and Chong.

Scott's demeanor was considered cool back then. At the office, many times loaded and glassy-eyed, he appeared to have the answer to the pressures of the times. That answer was, "Screw it."

As music director, he was in charge of searching out hip, groovy, cool music that we could get away with playing yet not blow off our core audience.

Occasionally, Bernie, Watson, and Drake would fly in to assist in the experiment.

On one of their visits to Fresno, they brought Paul Drew with them. Paul sat in on the decision-making because he was known as a radio wizard, just like Drake. Bill felt some loyalty toward Paul because back in 1960, they both worked at WAKE in Atlanta, Georgia, and when fame came to town for Bill, he made Paul one of his PDs.

They would hole up in a suite with a wet bar at the airport Hilton, and we'd all huddle around the radio, listening as we made changes to the format. Most of the time, I would be at the station so they could call me, and I could implement whatever the latest format tweak was. When Scott first stumbled into one of our hotel meetings loaded, you should have seen the looks on the faces of the Drake guys. I just knew they were thinking, who the hell is *that*?

The truth? As bad as Scott was, he wasn't much worse than I was.

On more than one occasion, my morning man, Dirk Robinson, would drive me around Fresno, on a Sunday morning, looking for where I might have left my Lincoln. I always tried to park it and hitch a ride if I got too buzzed. I was twenty-five years old and a driven, hard-working professional, but after dark, I turned into a wild man.

One Saturday night, I wound up at the Hi-Life Lounge, an upscale watering hole that was right out of the 'fifties. The golfers and Al Radka, a local radio and TV legend, went there for adult beverages. I was there with Cohoe, who had a pocket full of reds with him. Like an idiot, I gobbled a handful and washed them down with a couple of shots of scotch. Then with the courage and confidence I got from the drugs, I approached a beautiful girl sitting at the bar. Eventually, we left together, and she drove my Lincoln. *Cool*, I thought, *I don't have to drive. She'll get us to her apartment. I'll sleep it off and go home, sobered up, the next morning.* Didn't happen.

She pulled up in front of her house, got out of the car, and said, "See ya."

So much for that plan. I got behind the wheel and drove off. A few blocks away, I side-swiped a car. The sound sharpened my senses and told me that I was in trouble, and I'd better get out of there. About a block later, I ran a stop sign and broadsided another car driving down Belmont Avenue. Thank God, there were no injuries, but three cars were damaged, my Lincoln the worst.

On the way to jail in the paddy wagon, I thought, *well, you always knew you'd find a way to fuck up a good thing. You just pulled it off big time.*

The next morning, when I came out of my drug stupor, I looked up from my jail cell cot and saw Decker standing over me. He bailed me out, and on Monday, it was back to business.

I was convinced I'd be fired. How could they let me get away with this? Turns out it was another reason why some radio stations insist their DJs use aliases on the air. KYNO was never implicated in the newspapers. Ron Copeland was listed on the police blotter.

Gene took it well. I wasn't the first PD to do something like this in the Drake-Chenault organization. Before my incident, there was another Drake PD that did the same thing. He was driving drunk and also crashed a company car. They whacked him on the knuckles, and that was it. When I did it, Gene wasn't shocked. A few weeks later, he sent me back out to the dealership to pick out a replacement for the Lincoln.

This time, I chose a 1972 gold Chevrolet Chevelle with a white vinyl roof. I knew I had a second chance, and although the experience didn't slow my partying, it did make me more cautious about driving under the influence. Working for Drake had many rewards, and I wanted to continue enjoying them.

In 1971, Drake-Chenault put together an all-expense paid program directors gathering at Caesars Palace in Las Vegas. All of the PDs were in attendance. So were Bill Drake, Gene Chenault, Bernie Torres, Bill Watson, and a few record promoters who flew in to schmooze us. Since George Klein, the PD of WHBQ, Memphis, grew up with Elvis and was his best man in his wedding to Priscilla, we were all invited to attend an Elvis concert at the International Hotel.

Between 1969 and 1977, Elvis appeared before 2.5 million fans in 837 sold-out performances. Our seats were butted right up to the lip of the stage. It was almost like Elvis was sitting in our laps. We watched in awe as his neck scarves would sail over our heads to the girls behind us.

After the show, George led us to an elevator behind the stage that took us down to a room below. Milling around us were the Memphis Mafia, all wearing their diamond encrusted "TCB" (*Taking Care of Business*) necklaces. Once in this small meeting room, we all stood around a conference table and waited for Elvis. When he entered the room still wearing his stage outfit and dripping with sweat, he stood at the head of the table. There we were, in the presence of the king.

You might think Elvis would be bored with a situation like this, but that was not the case. He was a prisoner there, unable to leave his top-floor penthouse, and he welcomed any opportunity to interact. If he were to walk out on the casino floor to toss a few dice, he'd create a riot in no time. Instead of being bored with us, he was up, excited, and animated. It was obvious to me that he was high on something that elevated his energy, and his eyes were dilated and glassy. We asked him questions, and he graciously answered them. Someone asked him about karate, his latest passion.

I was standing near him at the time and to answer this question, he crouched down in a karate stance and started throwing air punches all around him, one of which came very close to my nose. I felt the rush of air as he brought his arm back to his side. I often wondered what would have happened if Elvis had accidentally hit me. He was known for giving away Cadillacs and motorcycles as if they were neck scarves.

One of the nights, we were all sitting at a Caesars Palace bar when in walked Jim Benson, who was an *independent* promotion man. He didn't work for any one record company; he worked for all of them at one time or another. He was a hired gun.

Every week of the year, a huge number of new singles flooded the radio stations. Only two or three got added to the playlist in a given week, when a record company felt a particular artist was going to be lost in the shuffle, they would sometimes seek a little extra help. Jim was as close as one could get to Bill Drake, and that night, he picked up the entire bar tab. Jim was a portly man in his late fifties who donned a sport coat and slacks and had a long, gray ponytail.

Late that evening as I was getting off my barstool to head for some much needed sleep, he walked up to me and said, "Hey Sean, put your hand out."

"Huh, okay," I mumbled.

"Take these and be in your room at 4 a.m." He dropped two $50 dollar poker chips on to my palm and walked away. In 1972, one hundred dollars was a lot of money. Off to my room I went. About an hour later, there was a knock at the door, and when I opened it—surprise, surprise. A scantily dressed Las Vegas hooker was standing there.

My bet is all of us were offered the same treat that night.

And that is all I am going to say about that.

My Lincoln Continental Mark 4

A KYNO business card

Joe Angel, Dirk Robinson, and Sean Conrad giving away one of the first Honda cars.

CHAPTER 11

"AFTER MIDNIGHT"

I arrived at KYNO in March of 1970, and three years later, I left Fresno for Chicago. During my last year or so at KYNO, I was approached many times by the general manager of KFIG-FM, the AOR station in town that was starting to make inroads in the ratings arena. Roger Turnbeaugh tried many times to get me to jump ship and join him at KFIG.

"Sean, come on over here and help me make radio history by beating KYNO," he would say.

"Roger, I love ya, man, but you can't pay me enough to cross over," I'd tell him. "I'm a hard core Drake-Chenault guy, and I got my sights set on KHJ."

Roger was raised on a farm, married to his high school sweetheart, and never cheated on her. He had movie star looks, a magnetic personality, and a strong drive to succeed. I wished he could afford to hire me.

Meanwhile at KYNO, we figured out we couldn't suc-
cessfully cross-breed AM radio with FM radio, discontinued
all tinkering with the Boss radio format, and got back to the
basics. Then with Gene Chenault's direction, we put an FM
station on the air and went head on against KFIG. The call letters were KPHD-FM. They were Gene's idea. He assumed the
listeners would be well educated and could relate to the
"PHD" reference.

He temporarily appointed me PD so I could create the
format, build the music library, hire a few jocks, and then hire
a PD to replace me. For a few months, I was programming
both stations. I found Paul Sullivan, my new program director,
in Long Beach, at KNAK-FM. I knew him from Detroit when
he was fresh out of college and was my gopher. I was always
impressed with his intelligence and knew right away, he was
the guy to take over the reins of KPHD-FM.

In the meantime, Roger was hired away from KFIG by
Alan Shaw, president of the ABC-FM Radio group, as General
Manager of Chicago's WDAI-FM. ABC owned and operated
seven FM stations in all the bigger cities like New York, Boston, Houston, Los Angeles, and San Francisco. They all programmed the AOR format. They were union shops and were
considered the big time.

One day in 1973, I got a phone call from Chicago.

I answered, and a familiar voice said, "I can pay you
enough now."

"Who's this?" I demanded.

"Sean, its Roger Turnbeaugh. You know I've always
wanted you to be my PD, and now I can afford to hire you.
How would you like to be the program director of an ABC-FM
station in Chicago?"

As PD of WDAI-FM, I would be replacing my buddy Big
Jim Edwards, who had held the position before me.

I hired Les Garland out of Milwaukee to replace me at KYNO. I always knew the day would come that I would outgrow Fresno. Les later went on to join Bob Pittman in New York City and helped launch MTV. Les is a good friend and a legend in the industry today.

So it was off to Chicago with Beth, the kids, and the cats. We both looked at the move as a fresh new start to our relationship. It was time to get out of Fresno.

We loved Chicago and found a house in Highland Park, about forty miles north of downtown. Highland Park was an upscale community with huge mansions everywhere. At the turn of the century, it had been a hunter and fisherman's paradise. The land had been dotted with hunting cabins because of its close proximity to Lake Michigan. The house we rented had survived demolition and had been added onto several times, to form a cute two-story, two-bedroom house. A few yards to our left, a few yards to our right, across the street, as well as behind us, were two- and three-story brick mansions that looked down upon our humble abode. We didn't care. It was a great place to live. It has since been bulldozed and has another mansion sitting in its place today.

Right after we moved to Highland Park in 1973, the locusts appeared. Cicadas are two-inch long with prehistoric features. They only surface every seventeen years. They would not appear again until 1990, and in 2007, and wouldn't again until 2024. Once they emerge, they spend their short two-week life climbing trees, shedding their skins, and reproducing. Then, they go back underground to hibernate until the next time. They can number several hundred thousand per acre, and were everywhere the eye could see that year in Highland Park. Walking from the car to the front door, I could not help but walk on hundreds of them as they crunched underfoot. There was concern that the sound made by the cicadas would drown out the music from the Highland Park outdoor music festival with the Chicago Symphony Orchestra in 1990. The schedule was adjusted to correspond with their presence. It was exciting, yet creepy living among them, but Beth and I were happier than we had been for a long time.

Every weekday I'd walk two blocks to the Chicago-Northwestern Train station and settle in for the one-hour ride to the windy city. Wearing a suit and tie, with hair down to my shoulders, I'd suck on a cup of coffee and prepare for my day.

Once the train arrived at the terminal, I'd walk across a bridge to the other side of the Chicago River, get on a boat, and finish my commute to the Stone Container Building just across from the Wrigley building on Michigan Avenue. Every evening, I'd reverse the journey and head home. Many times, I'd look up at the skyscrapers as I walked around town, thinking how different this Chicago experience was compared to seven years before during my six-week stay in 1966 at the YMCA Hotel studying to get my First-Class License.

In 1969, the morning man on WKNR, Detroit was Mark Allen. The afternoon drive jock was Sean Conrad. We were good friends and simulated a rivalry when we were on the air for entertainment value. In 1973, Mark Allen became Bob Dearborn and moved to Chicago to be a jock on WCFL. In 1973, Sean Conrad was also in Chicago as PD of WDAI-FM. You already knew the part about me but I have a reason for detailing out the information about Bob Dearborn.

When I realized one day that Mark Allen had become Bob Dearborn and was in Chicago, I tracked him down. Bob had no idea I was also in Chicago because Roger Turnbeaugh had talked me in to shedding my air name since I was not going to be on the air at WDAI, and go back to using my real name of Ron Copeland. I called WCFL and left Mark/Bob a message to call Sean Conrad who was now Ron Copeland. So begins an evening right out of the *Twilight Zone.*

Bob and I decided to get together and toss down a few for old time's sake. He gave me the address of his downtown Chicago apartment, and one Sunday afternoon when I was in town doing something at the station, I took a cab to his place. We planned on having a few drinks at a local watering hole and talking about our time in Detroit. Well, Bob was always a J&B Scotch man and, several bottles later, around 11:30 p.m., I realized it was time to go home. I was also aware that the last train to Clarksville, I mean Highland Park on a Sunday night, was at 12 midnight. If I was going to be on that train, I had to dash. With the few dollars I had left, I hailed a cab which got me to the train station at 11:50 p.m. As I ran through the empty, cavernous train station, which was right out of a movie, I let out a sigh of relief because I had made it on time!

When I got to the station platform, there were four or five trains on the tracks, idling away, as if they were getting ready to pull out of the station. After all that scotch, it all seemed so surreal because there were no people anywhere, yet these trains were belching smoke and fumes and all lit up. Oh well, I thought and climbed aboard my train, took a seat, and quickly fell asleep (passed out). Around 2 a.m., I awoke and realized the train had not moved an inch. What the hell! I got off the train, back in to the eerily empty train station and walked to the window where you buy your tickets. I explained to the ticket seller that I was under the impression the last train to Highland Park was at 12 midnight. What was the problem? Why was the train just sitting there?

"Sir, that's the case every night except Sunday night, when the last train is at 11 p.m.

Oh shit! I had no money. No cell phone, of course. No credit cards. What the hell was I going to do? I couldn't get on the boat on the Chicago River and have it take me to work like I would normally do. There were no boats running at 2 a.m. on a Sunday morning in downtown Chicago. Besides, even if there were, *I had no money* on me. For that same reason, I couldn't hail a cab either. There was only one thing I could do and that was to walk to the Stone Container building and sleep on the sofa in my office. The exact distance from the Chicago & Northwestern Train Station at 500 West Madison Avenue to the Stone Container Building at 410 North Michigan Avenue is 1.8 miles. At 2 a.m. that is a very long walk. I stumbled and weaved my way through canyons of classic old Chicago brick buildings and sky scrapers, never knowing what was going to be around the next corner.

I eventually got there somewhere around three in the morning. I was never so happy to see that worn out old sofa in my life. I will never forget that night. There is no other way to explain what it felt like walking through dark side streets in downtown Chicago after midnight on a Monday morning when there's not a person in sight, except to say that it was like being in...(do-do-do-do, do-do-do-do)...*The Twilight Zone*!

During my short tenure as PD of WDAI, I had the pleasure of working with some great people; Jin Kerr, Dave Van Dyke, Greg Budil, Mitch Michaels and Roger Turnbeaugh just to name a few.

When I first started at WDAI, our morning man was Jim Kerr, who was one funny guy. We figured he would be even funnier with another guy sitting next to him. I knew I needed someone who would stand out and maybe cross the line now and then. There were no Howard Sterns yet. Enter my friend, crazy Steve Randall.

"Steve, pack your bags. You're coming to Chicago," I said. Later in his career, he was known in the Seattle area as

Crazy Steve. He and Jim were great together, but the experiment ended way too soon.

He'd have more opportunities at stardom down the road, but in Chicago, after three months, personal problems in his hometown of Fresno forced him to move back. He was re-hired at KYNO by Les Garland.

Toward the end of my fourth month at WDAI-FM, Roger and I were busy trying to elevate the station's low ratings. I was having great fun as PD and figured I'd be in Chicago a long time. In the meantime, Charlie Van Dyke, was doing mornings on WLS, our AM station, located on the floor above us. WLS stood for World's Largest Store because Sears owned it. Van Dyke had been the morning man on CKLW, before coming to the powerful and legendary 50,000-watt AM station. One day Charlie came in to my office to make an announcement.

"Sean, are you sitting down?" he asked. "You are looking at the new morning man for KHJ, Los Angeles."

"Get out of here," I said. "My life's dream is to go to KHJ. So, when do you start?"

"Well, that's the hitch. I have a contract here at WLS and can't get out of it for another three months. I wish I could go now, but that's just the way it is."

The following week, a phone call awoke me at six on a Saturday morning.

"Is this Sean Conrad?" the caller asked, in a nasal voice.

"It is. Who's this?" I replied with a little disdain because someone had the nerve to call and wake me on a Saturday morning.

"This is Paul Drew. How would you like to be the PD of KHJ?"

The famous ABC FM logo

Big Jim flashes the bird while sitting in his WDAI office, the same office I
would occupy as his replacement a few months later.

With Steve Randall and his newsman, Ron Hill,
during his morning show at WDAI

Greg was my talented, young research director who
later made it onto the radio! He's now a successful
talk show host in Montgomery, Alabama.

CHAPTER 12

"CALIFORNIA SUN"

This had to be a joke. Yes, it was true that there had been a shakeup at RKO with Drake-Chenault exiting and Paul Drew, who was the PD of KHJ at the time, chosen to replace them. Paul had called to hire me to take over his old job. I was disappointed that I wouldn't be going to KHJ under Drake-Chenault, but there was no way I could turn down the opportunity.

A few days later, I flew in to LA to be interviewed by Tim Sullivan, the General Manager of KHJ. I was hired even before I returned to Chicago.

Then I had to go break the news to Roger. It was one of the most difficult things I've ever had to do. Roger and Allen Shaw both tried everything they could think of to get me to stay, but there was just no way I could pass up this opportunity. KHJ! By golly, I did it. I'd be the PD of KHJ. Now it was time to tell Charlie.

"Hey, Charlie, you busy?" I asked on the intercom.

"Not right now. I just got off the air. What's up?"

"Uh, could you come down here for a minute? I want to talk to you about something."

"I'll take the first flight of stairs I can find and be right there."

Charlie got to my office in less than five minutes.

"Sit down," I said. "I have something to tell you."

"Okay, shoot."

"Guess who called me last week."

"Elvis Presley? John Lennon? I give up. Who?"

"Paul Drew."

"What? You're kidding me. Why?" he asked.

"You are talking to your new PD."

"Oh my god. No! Really? C'mon, you're pulling my leg."

"It's true, pal, and I can get out of here with two weeks' notice."

"You mean, you'll be there in two weeks, and I got to stick around here for another two months?" he asked. "Somehow, that's just not fair."

ᏸᏰᏸ

I headed to LA with Beth, two kids, and two cats. RKO knew how to treat its new PD. A moving truck pulled up to our house, loaded up our furniture, and headed west. They had my Chevelle picked up, taken to the train station, loaded on to a flat bed, and transported to LA. The four of us were flown first-class to LAX, where a limo picked us up and took us to the Wilshire Hyatt House, all expenses paid. We stayed in a suite at the hotel for a full month while we looked for a home to buy in the San Fernando Valley. Soon we were living in our new three-bedroom home with a swimming pool.

A few months later, Berry Gordy physically re-located Motown Records to Los Angeles. That meant my buddy, Russ, and his family would be moving with them. Amazingly enough, they bought a home directly behind ours only separated by a cinder-block wall.

And there I was, a first-time homeowner, driving a 1969 Porsche 911T to work every day. I can't explain the thrill of flying down the Hollywood Freeway to the Gower exit, looking over at the Capitol Records building, the one that looks like a stack of records, and then flashing my badge to the guy in the guard shack of the KHJ parking lot.

KHJ was located at the corner of Melrose and Gower in Hollywood and was smack dab in the middle of where all the action was. The back wall of the building that housed it was butted up against one of the west walls of Paramount Studios. As program director, my desk was only a few yards away from where the Marx Brothers, Elvis Presley, Bob Hope, and Alfred Hitchcock crafted their art for the cameras. Just across the street, was the studio where the Beach Boys recorded "Surfin Safari" and many of their other early hits.

One of the first promotions conducted at KHJ just after my arrival in July of 1973 had been put together by Drew before I got there. With Charlie Van Dyke due to arrive in about eight weeks, Drew wanted to start building the morning drive show's audience in advance. He arranged with the record companies to have some of their hottest recording artists come and sit in with Bill Wade, who was holding down the slot for Charlie. In a seven-day period, Diana Ross, The Carpenters, Donny Osmond, Dionne Warwick, David Cassidy, Jerry Reed, Glen Campbell, and Mac Davis all appeared live with Bill Wade. My first experiences at KHJ included shoulder-rubbing with some of the hottest stars of the day. Rhonda and Robyn, my little girls, got to go to work with their dad on the days David Cassidy and Donny Osmond were there.

Radio heaven quickly turned in to radio purgatory. I soon realized that Paul Drew was even more difficult to work for than I could have imagined. I was required to be on call twenty-four hours a day, seven days a week, literally. Paul demanded that every day I vary the time I arrived at work. Several days a week, he insisted I arrive at 3 a.m., just to keep the employees on their toes. He wanted me to bust them doing something he thought they shouldn't be doing. I had the nerve to take a three-day weekend after six months of working seven days a week. Drew reluctantly said okay, but I had to give him the name of the motel we were going to be staying at in Morro Bay. Since we didn't have a phone in our room, the last thing I expected was to have to talk to him. The second day, Drew called the motel manager and insisted that he go to our room, knock on our door, and make me come to the phone in the middle of the night. He then demanded that I return to the station for some manufactured emergency. The first home answering machines were just hitting the market around then, and I was one of the first to buy one so I could hide from his controlling self.

I had acquired four tickets to see the taping of *Old Blue Eyes Is Back*, the now famous TV special featuring Frank Sinatra. I had rented a tuxedo, and Beth had bought an expensive dress. She had to go without me. Drew demanded that I spend the evening at the station keeping an eye on things while his wife, Ann, completely re-typed the program log. It was insane, and I regret to this day not telling him and his toad, Harvey Mednick, to shove it. Since Drew's office was physically located in another building, Mednick was his "spy" and was heard on the phone with Drew stabbing me in the back on numerous occasions. What really sucked was that as I sat at my desk twiddling my thumbs that night, the Sinatra taping was taking place right next door at Paramount Studios.

Nothing was good enough for Paul Drew. I was in the production room while Charlie was cutting station IDs one day. He'd record a few and then call Drew and play them for him over the phone. Ninety-nine percent of the time, Charlie had to re-do them.

"Watch this," Charlie said to me one day. "Hello Paul, here are the latest station IDs you asked me to do. Can I play them for you now?"

"Why else would you be calling?" replied Drew. "Go ahead."

Charlie played them.

"I want you to do them over," Drew said. "There's not enough energy. Do them all over and get back to me."

Charlie never re-recorded them. He just waited twenty minutes and called Drew back.

"Okay, Paul, I re-did them," he lied.

Engineer Jon Badeaux played them back over the phone.

"That's much better," Drew croaked.

"Thanks, I've been up since three a.m. Can I go home now?" asked Charlie.

In December of 1973, RKO sent all us PDs to the Coronado Resort in Puerto Rico. It was a company sponsored rah-rah event that included all the station GMs and Sales Managers. There were to be meetings during day and gambling, partying, and chasing women in the evening. Except that I never got the chance to get down and get funky with my coworkers because Drew pulled me out of the second day's meetings and walked me to the lobby.

"Sean, you need to get to the airport and get back to LA as soon as possible," Drew commanded.

"Why, Paul? I just got here yesterday."

"I don't care. We have an emergency back home and you need to be there."

"Okaaaay, what kind of emergency?" I asked.

"I just found out there's a station in town that's switching to top 40," Drew said.

"Well, how can I possible stop that from happening?"

"You can't. But you can be there to make sure we're sounding tight and bright, and the jocks are on their toes!"

The absurdity of this control freak's ridiculous command was obvious. We were both going to return to Los Angeles in less than forty-eight hours anyway. There was absolutely nothing that would benefit KHJ by my premature return. Off I went in a cab to the airport for the eight hour flight back just to make sure the jocks were not screwing up.

My jock staff at KHJ included Bill Wade 9 a.m. to noon, Danny Martinez, noon to 3 p.m., Barry Kaye, 3 to 6 p.m., Captain John Lodge, 6 to 9 p.m., Bobby Rich, nine to midnight, Johnny Williams, midnight to 6 a.m. and of course Charlie Van Dyke, 6 to 9 a.m.

There were good times at KHJ. Once we conducted an on-air contest where the winner got to spend a day on the town with Charlie Van Dyke. Radio stations always promoted the morning man because if they could get the listeners tuned in, in the morning, they might listen all day. The winner and his wife were to be picked up in a limo by Charlie and his wife, Adele, and whisked off to Disneyland and then to dinner at the Brown Derby. Beth and I tagged along.

We entered Disneyland in the limo from a staff-only gate. We were shown what goes on behind the scenes and had lunch in what was Walt's private dining room. When we went around the corner at the Pirates of the Caribbean, there was a door that most people would never notice. The Disney tour guide unlocked the door and walked us in to a glass elevator that appeared to be from the turn of the century. We all piled in, and it took us to the second floor where, once the doors opened, we were in another world. Every wall, the floor, and the furniture had a story behind it and originated from some other part of the planet. It was totally elegant. They told us that when Walt was still alive, he'd have lunch there every day, looking out the one-way windows, marveling over what he had created. I could almost feel his presence in that room.

Then, we piled into the limo and headed to the famous Brown Derby Restaurant. The restaurant itself was in the shape of a...well, a brown derby, and it was an institution in Hollywood. Again, we were treated like royalty in the land of make believe.

છ⁄ళઝ

I got a call a week before Christmas 1973 from Marshall Blonstein, a guy I had known as far back as KYNO. He was co-founder of Ode Records and he was calling to invite Beth and me to a Christmas party. Not just any party. This one was being held at the Hollywood Hills home of Lou Adler, who was Marshall's partner at Ode and the producer of such groups as Jan & Dean, The Mama's and Papa's, Johnny Rivers, The Grass Roots, and Carole King. He produced the movie *Up in Smoke* with Cheech & Chong, and was cutting edge cool in the '70s.

"Sean, don't miss this party," he said. "You'll have a great time. There will be many stars there, so be prepared to be blown away. I'll see you there."

We drove to the party in our newly acquired Porsche 911T, which was parked for us by a valet. In typical Holly-wood fashion, these weren't your average, run- of-the-mill valets like the ones that park cars at fine restaurants. These guys wore tennis shoes, vests with no shirts, bowties, and top hats. We slipped ours a few bucks and started walking towards Lou's luxurious home, which was elegantly decorated for Christmas.

Just outside the front door before we entered the house, we were offered smoked herring hors d'oeuvres and a shot of tequila to wash them down with. After that, it became very dream-like. Goldie Hawn in that football jersey. NFL star Jim Brown, Warren Beatty, Sonny & Cher, and of course, John Lennon.

And yes, I can say, all these years later, I met John Lennon. It would be years, however, before I got off the roller coaster and straightened out my life.

A little later that night, we wandered around the house, and soon noticed that other guests were going up and down the stairs. We decided to check it out. As we entered a room, which was reeking with the smell of marijuana, a very young girl, who was maybe twelve to thirteen years old, handed us a big fat joint. Beth struck up a conversation with her and found out she was the daughter of the actress Britt Eckland and her ex-husband Peter Sellers and was now living with Adler.

The little girl, whose name was Victoria, told us, "My daddy says this stuff makes you goofy."

We fired up the joint and passed it over to Johnny Rivers, who passed it to Britt, who passed it down the line, as "the room where you go to get high" started filling up. (No, Victoria did not imbibe.) The rest of that evening is unfortunately clouded in a haze of tequila and grass, as were many evenings at that time in my life.

Not only did we get invited to parties because of KHJ, but my close connection to Russ and Motown generated many other perks.

"Hey, Sean, wanna go to a party?" Russ asked.

He was the recording engineer on many Jackson Five albums.

"When have I ever said no to a party? Who? What? Where? When?"

"At Joe Jackson's house out on Havenhurst, in Tarzana. It's a family kind of a party so we can take our wives and kids too."

"Is the whole Jackson Family going to be there?" I asked.

"Yep, everybody. It's a casual barbecue out by the pool. We just finished recording the Dancing Machine album, and this'll be a little celebration."

It was a warm, summer southern California afternoon when the eight of us, Russ, Joanne, Beth, and me, along with our kids, pulled in to the driveway of the Jackson Five's home. At the time, Michael was only sixteen and he, along with his brothers, had already gone way beyond super stars. We walked from the car to their backyard, which was surrounded by a grove of trees.

All eight of us converged on one table and kind of headquartered there the rest of the afternoon. The yard was all tricked out for a party. There were large, round tables scattered all around a pool that had two swimming dolphins painted on its bottom. The whole family was there, including their mom and dad, Joe and Katherine, along with record company dignitaries, family friends, support staff, and many kids romping and playing around the pool and in the game room. One of those kids romping around was obviously Janet Jackson, who would have been seven years old at the time. The Osmonds were there, too. If you check your rock and roll history books, you'll find that Michael and Donny became pretty good friends around that time, so it was no surprise to see Donny. All four of our kids got their pictures taken with Michael.

Later in the afternoon, Redd Foxx sauntered in walking the way only Redd Foxx could, wearing a red beret. Then he and Hal Davis, who produced the *Dancing Machine* LP, joined us for a while at our table. Hal was a funny guy who loved to tell a good joke. When they sat down, they were drinking what looked like glasses of milk.

"Hal, what the hell are you guys drinking there?" I asked. "Milk?"

Redd replied, "Yeah, we drinkin' milk all right. Mother's milk."

"You mean your mother came to the party, too?"

Hal cackled. "No, this is scotch and milk, plus a little sweetener thrown in for good measure. They go down easy and kick your ass."

Soon, we were all drinking Mother's Milk.

Then Beth had to go to the bathroom. Hal showed her to Michael and his brothers' bathroom. There was a glass door that separated the shower and toilet from the rest of the bathroom, but it was broken because their little brother, Randy, had fallen in to it and cut his leg a week earlier. He was on crutches at the party. So, Hal left and Beth sat down to pee. A moment later, she looked up. Michael was standing there watching her.

Beth mumbled something about did he have to use to the bathroom, and in his soft, childlike voice, he said, "No" and quickly left.

Shades of things to come? I had no idea then, and I have no idea now. At the time, enjoying the hospitality of the Jacksons was a high point of my professional life.

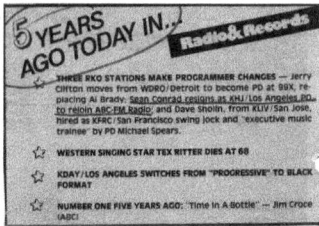

5 years ago today

Interviewing Rodney Allen Ripey

With Paul Drew, music director,
Meredith Lifson, and Mac Davis

With John Long, George Klein,
and Gerry (Cagle) Peterson at
Puerto Rico RKO meetings.
One year later I had lost 65 pounds!

Rhonda and Robyn meet
David Cassidy at KHJ

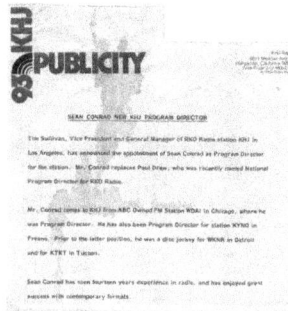

KHJ Press Release

KHJ Business Card

Partying around the Jackson pool

Michael Jackson milling
around the party

My daughter Rhonda with Randy
Jackson who's holding his crutches

Terry Terrana, Robin, and
Michael Jackson

Terry Terrana, Michael Jackson,
and Rhonda

Seated in the background:
Katherine Jackson

Joe Jackson in the background,
Rhonda in the foreground,
looking for more face time.

KHJ Lobby

93-KHJ, 5515 Melrose Ave,
Hollywood, California

With Paul Black (Tree) from Columbia Records,
recording artist David Essex (Rock On),
and Terry from Columbia

CHAPTER 13

"CLAP FOR THE WOLFMAN"

By the first quarter of 1974, I'd had it with Drew and KHJ. So I quit. It was so bad there that I figured stepping in to the unemployment line couldn't be any worse. Something would come along, and, besides, Beth and I had managed to save some money.

One night we were having a pool party with friends and we all decided to drop some acid.

Party, party, party. Ring, ring.

"Beth," I managed to say, "Don't answer the phone. I don't want to talk to anybody in this condition."

Too late. She already had the evil phone in her hand.

"Hello. Uh, I think he's here. Hold on a second, I'll check." She put the palm of her hand over the phone and said, "It's a guy named Fred Constant. He wants to talk to you. He said he was calling from Hawaii."

"Hawaii? Shit, I'm stoned out of my mind." I didn't want to do it, but curiosity got the best of me. I went ahead, took the call, and introduced myself to Fred Constant, the owner of KPOI Radio in Honolulu.

"I recently bought the station," he told me. "I need a program director. I called Gene Chenault and asked him if he could recommend anyone. He gave me your name and phone number. Have you got a few minutes to talk?"

Oh no. I was tripping through outer space, and this guy wanted to talk serious business with me.

"Sure, I've got a few minutes," I said.

For the next hour and a half, I was quizzed about what I would do to resurrect what had once been a dominant radio station in Hawaii. The station was seriously suffering from years of neglect and was way down in the ratings. Frankly, I have no idea what I said in that ninety-minute interview. I did know right away that he had called Gene, then me, because he wanted his new radio station to sound just like KHJ. So I went on auto pilot and just parroted everything I had learned during my days as a Drake PD. He said goodbye, thanked me for my time, and asked me to call him in a few days. I was dripping with sweat when I got off that call.

For the next few days, we started fantasizing about what it would be like to live and work in Hawaii. Neither of us had ever been there, but we'd seen enough Hawaiian sunset pictures to know this move could be our greatest adventure yet.

* свсэ*

Fred was tight with money, and my paycheck proved it. Even in 1974, his two grand a month wasn't enough money to live comfortably, and there were no record hops to up the ante. The people of Hawaii had never been exposed to a jock playing records on a high school gymnasium floor. But our desire to live there overpowered the reality of the situation, and I ended up working twelve hours a day.

Before Fred bought the station, high-rise condos had started popping up all around the radio station towers. Gradually, the underground copper grounding wire that was so important to a good signal started getting ripped up by bulldozers as they built the condos. Then as each twenty- to thirty-story building grew higher and higher, the signal got weaker and weaker. It doesn't matter what you're programming if half the population on the island can't hear you because of signal strength. Okay, so it was my fault for not checking out those kinds of issues before dragging my family and all our belongings across two thousand miles of water to a land-locked island.

None of that stopped me from absolutely busting my ass to make that station sing. My first duties were to trim back the on-air staff. The previous ownership had allowed the staff to balloon up to twice as many announcers as were needed. It was no fun letting people go, but I had no choice. The staff I kept was the best of the crop, and together, we brought the ratings up, and we did it with a tiny promotion budget.

Fred wanted the moon but didn't want to pay for the fuel to boost the rocket, so I improvised. Using the principle of trading services, I managed to get my hands on vacations to the other islands, meals at local restaurants, karate lessons, anything of value for on-air contest giveaways. We made that station sound like we were giving away the world. The record companies supplied records, T-shirts, and other rock and roll paraphernalia for us to give away.

Every Friday through midnight Sunday, all during the Solid Gold Weekend, we'd air some sort of contest, which had to be manufactured out of thin air. When the record "Kung Foo Fighting" by Carl Douglas came out, our jocks would say,

"Be the fifth caller when you hear this sound, *aiyaiii KPOI.*"
It was the sound of Jeff "Boogie Bear" Kauffer throwing a
karate chop. The winner would receive karate lessons from the
Waimanalo Karate Institute. We'd trade out the lessons for
advertising.

Another weekend, each Beatle had a solo record in release. I got fifty copies of each single from the record companies, and every hour we'd announce, "KPOI puts the Beatles
back together. Be the sixth caller when we play a Beatle's
song and win a Beatle four-pack."

Nice promotion and it didn't cost Fred a nickel. Coming
up with a new and different promotion every weekend required
tremendous creativity, and with input from my loyal on-air
staff, we managed to do it for an entire year.

Another major promotion we put together was one I had
pulled off at KYNO, the most popular high school contest. The
high school that turned in the most sheets of paper with the
words *Vote KPOI* handwritten one hundred times per page
would win a live rock concert at their high school starring Hawaii's most popular local band, The Society of Seven. Kids
drove up with truckloads of boxes of entries. The promotion
turned the town upside down. They went nuts for this contest.
High schools competing with high schools always fires up
some sort of primal instinct, and that desire became infectious.
The night before the end of the contest, Riley Cardwell, our
seven-to-midnight jock, answered the request line.

"KPOI request line, this is Riley Cardwell. What can I
play for you?"

"Riley this is Jim Stratton, principal over here at Roosevelt High School. You need to stop by the gymnasium when
you get off the air tonight and witness this."

"But it'll be after midnight," he said. "What for?"

"Believe me, you need to see this."

So even though he had just spent five long hours on the radio, on his way home, Riley decided to stop by Roosevelt High at 12:15 on a school night and see what all the fuss was about. In that gymnasium, a good twenty-five cafeteria-style tables were set up with high school kids seated all around, writing, "Vote KPOI" on thousands of sheets of paper. Hundreds of boxes were stuffed with reams of paper everywhere.

The next morning, five minutes before deadline, a caravan of cars pulled into the KPOI parking lot and dropped all those boxes off for us to count. Roosevelt won the contest turning in over a million hand-written *Vote KPOI*'s. We didn't count each one. We came up with a system to weigh each box and assigned a number of sheets to that weight. Again, we'd pulled off an incredibly successful promotion that cost Fred nothing.

On Saturday nights from 9 to midnight, we aired the syndicated *Wolfman Jack Show.* The real one this time. One weekend, there was going to be a Guess Who concert at the H.I.C. Arena, with Wolfman appearing on stage performing "Clap for the Wolfman," a huge hit for the Guess Who. A few days before the show, we took the entire group and Wolfman to lunch at the Tahitian Lanai Restaurant a few blocks away in the Ala Moana Hotel. We had a trade there, of course. It was a fun time eating poi and other Hawaiian foods.

Toward the end of the meal, Wolfman took me aside.

"Hey, Sean, you think you can get some of that super fine Hawaiian weed for your old buddy Wolfman?" he asked. "I can't tell you how much I'd appreciate dat. You understand what I'm saying, Sean?"

"Sure, man, I know where to get some Thai Stick that will paralyze you."

"Oh, man, I can tell you that would make da Wolfman one happy son-of-a-bitch."

I went to a friend's house and bought a couple of sticks that I delivered to Wolfman's room in the Kahala Hilton Hotel. The Wolfman was a big dude, probably six foot one and weighing in at two thirty five. At the time, I weighed 170. He grabbed me in a bear hug, or maybe it was a wolf hug, and squeezed the living shit out of me.

"Oh, man, thank you, thank you, thank you! Da Wolfman's gonna have fun tonight, baby."

That night we put Wolfman on the air, live in the parking lot in the Poi-mobile, a converted mail truck that had a multicolor paint job and a functional on-air studio inside. Hundreds of listeners stopped by in the rain to pay homage to this radio legend. Halfway through the show, the Guess Who pulled up and joined in on the air. After all, everybody knew the Wolfman from his appearance in *American Graffiti*. Cars drove by with high school girls yelling out the window, "We love you, Wolfman!"

If you want to hear some of that magical event, go to YouTube and type in *Wolfman Jack on KPOI*.

ℯↄℯↄ

In the summer of 1974, a concert promoter came to us to talk about a rock concert he was planning to bring to the island. His intention was to duplicate the success of Don Branker's Cal Jam One earlier that year. The only reason Don himself wasn't there to help produce Hawaii Summer Jam was because he was in Twin Falls, Idaho, producing Evil Knievel's famous failed attempt to jump the Snake River in a Skycycle. In his place he sent his partner Rich Linnel to produce the show for the concert promoter.

Hawaii Summer Jam was no Cal Jam, that's for sure. It took place on Saturday, August 31, 1974 at Hawaii Raceway

Park. On the bill, were Billy Preston, War, Black Oak Arkansas, Brownsville Station, and local stars, Cecelio and Kapono. KPOI Radio was the presenting station, and Beth and I both became involved in its production. I did all the radio commercials, gave away tickets on the air, and heavily hyped all the groups with increased playlist rotation. Beth went to work for the promoter as a bookkeeper and was in charge of ticket pre-sales, so she was witnessing it from the inside. From the beginning, she became suspicious because at the end of each day, she was instructed to deliver a bag full of money to a local car dealer, who put it in a safe. We knew something wasn't right.

Whenever possible we took the girls to any rock concerts we attended, and that day was no exception. Since we would be working, we also brought a babysitter along to keep an eye on them. Not long after the concert started, it became apparent that nowhere near as many people as expected would be there. Of course, that meant less money in the pockets of the promoter.

Just then, three long stretch black limos had quietly pulled in to the backstage area, and out of each one came four humongous local-type dudes wearing Blues Brothers dark sunglasses. They walked to the cars trunks, opened them, and pulled out rifles.

Beth was in the motorhome office back stage when a band manager came in, one side of his face swollen and red.

"What happened to you?" Beth asked.

"One of those guys told me we were going to be paid only half the money we were contracted for. I told him screw you, we aren't going on then. That's when he knocked me down."

Not long after that, I saw all of these gorillas chasing two long-haired roadies for Black Oak Arkansas across the backstage area, catching them, and beating the holy shit out of them.

"Beth, we have to get the girls out of here quickly," I said.

We told the babysitter to take Rhonda and Robyn to our car and leave immediately.

We stayed out of harm's way and close to the stage area in view of the audience, who had no idea what was going on behind the scenes. The concert went off as planned, but we found out later that each group was told they were only going to be paid half, and the goons kicked the shit out of anyone who complained. One of the roadies from Black Oak spent three months in the hospital. We also heard later that as each group finished their sets, they headed straight for the airport. Apparently the concert had been funded by the underworld in Hawaii, and they weren't going to take a loss under any circumstances.

Honolulu is a close knit community. There was never a policeman in sight and nothing in the newspaper in the days that followed. Little comes up in an online search, and as far as I know, this is the first time it has ever been written about.

೧೧೧

One day, while at KPOI, I got a call from Bill Drake, AKA Phillip Yarborough.

"I heard you were in Hawaii," he said in his unmistakable voice. "How's it going?"

"I'm working my ass off trying to make KPOI a contender again. Thanks to everything I learned from you, I seem to be doing it."

"We're thinking about flying over there, and I wondered if you have any connections at hotels. My wife and I need a vacation from Los Angeles."

"We do Bill, at the Kona Hilton on the Big Island. Let me see what I can do."

Living and working in the islands afforded me special deals and discounts. I was considered a *kamaiana,* which means a local in Hawaiian. I used those special privileges and radio station connections to secure Bill a super deal. I was honored to do it for him.

"Bill, here's a contact name and phone number at the Hilton. Give them a call. They'll take care of you."

"Thank you, thank you, a million thank yous. Hey, why don't you and your family join us? I'm sure you could use a little break too."

Let's see. Hanging out with the big guy for a week, shootin' the shit, talkin' radio, and sucking up umbrella drinks?

"Hell yes!"

"Great, Sean," Bill said. "I can't wait to hear your radio station. See you in a couple of weeks."

We had a wonderful time that week smoking cigarettes, slathering Hawaiian Tropic cocoanut-scented sun tan lotion on ourselves, and cooking in the hot sun.

ↂↈↂ

In Honolulu during the 'fifties, 'sixties, and 'seventies, competitive radio station KGMB had a morning man who owned the listening audience. His name was J. Akuhead Pupule, which meant crazy in Hawaiian, and the locals loved this guy. His real name was Hal Lewis. He was Jewish and from the Bronx. Somehow he connected with the people of Hawaii. Steve Randall, my buddy from KYNO and WDAI, was my morning man at KPOI. When he departed for health reasons, I wanted to come up with my own version of Akuhead and remembered an old friend from KYNO. His air name was Big John Carter, real name John Yount.

John was truly an eccentric. He was grossly overweight and had a sense of humor that knocked me out. Even though he had high ratings wherever he went, no one would touch him because of his temperament. It wasn't drugs. He didn't smoke or drink. Yet, despite his talent, he couldn't conform to any structure long enough to hold a job. If he wasn't fired within the first year or so, he would quit. John had inherited some money and didn't need an income to survive.

I called him in Fresno, where he was currently living and said, "Hey, Big! Have you ever dreamed of living on a Pacific island, enjoying ocean breezes, and looking at beautiful women in bikinis on white sandy beaches?"

"Okay, what's the joke, Sean?" Big asked.

"I have an opening for a morning man, and I want you."

"Well, you know I would only do it if I can get away with pushing the envelope," he said. "You know me. I don't want to get you in trouble with the FCC."

"John, that's exactly what I want. I want you to take chances and get me some listeners. You'd have my permission to do whatever you want to do to put a dent in this Aku guy."

Big John was well aware of the legendary Honolulu morning man and understood exactly what I was looking for.

Fred Constant lived on a fifty-foot sailing yacht that was tied up near the mouth of the Ala Wai canal not far from the station. It became one of our party places. We'd hold staff meetings over beer and hot dogs once a month. At one of those staff meetings, we were slamming down a few just before Big John arrived on the island, and we decided he needed a more powerful on-air name. A name the locals could relate to.

After a few drinks, Fred blurted out, "How about we call him "Maxx Mahi Mahi?"

Big John loved it, and when he went on the air with that name, he was soon the talk of the town because it was an obvious declaration of war on Aku.

There had just been a hotly contested gubernatorial election in Honolulu in which a local, George Arioshi, had defeated mainlander Randolph Crossley. Emotions were high as people took sides on an election that had been decided a month before.

One morning I was getting ready for my drive over the Pali Outlook to Waikiki from our beach cottage in Kailua, listening to Maxx like a good Paul Drew-trained PD would do.

"Good morning, this is Maxx Mahi Mahi," he said. "It's ten past eight on this beautiful December morning, and I have a special announcement to make. I just checked the news wires, and it looks like there was a computer malfunction in the vote-counting process. Randolph Crossley has won the election, not George Arioshi."

After pulling a papaya off the tree in the backyard for breakfast, I headed to the station. The news story Maxx kept talking about meant nothing to me. I wasn't into politics back then. But when I walked in to the station, I was accosted by general manager Steve Wrath almost immediately.

"Sean, come into my office please," he said. "Did you know about this?"

"About what?" I asked.

"This phony news story Maxx is hammering to death on the air this morning."

"I never knew a thing about it. Wait, you mean it's not true? He made it up? He never told me he was going to do it."

"Get him off the air. Put Glen on early and get him in here so we can talk about this."

It was a big deal. The FCC had recently started cracking down on stunts like this. We didn't need them breathing down our necks or fining us.

John/Maxx sat across from me on a chair too small for his girth. His face was red, and he pouted like a little kid.

"Maxx, why didn't you tell me about this first?" I asked him.

"Well, because you would have said no."

"Look, Maxx, I have no choice but to fire you," Steve announced. "But if we get any listener response in your favor, we'll just tell everyone we decided to suspend you for a while instead."

Between December 4 and December 7, the Honolulu Star-Bulletin, the Honolulu Advertiser, the Associated Press, and Billboard Magazine ran numerous stories about the incident. Headlines like, *Election Day Gag Mis-fire* and, *Maxx Mahi Mahi Fired By KPOI.* Then a few days later, *Mahi Mahi Still Off The Air.* Even Aku was talking about it on his show.

It was a promotion dream come true for the station. And it cost Fred nothing, except having to make nice with the FCC, which he did. The calls started coming in from listeners who thought it was not right to fire Maxx. They all parroted, "It was just a joke, and we want our Maxx back."

We milked it for a few weeks, playing listener calls on the air to fuel the fire. Then, we put him back on the air when the positive effects of the stunt started to peak.

"I warned you when you hired me, Sean," he said.

"I know, I know. Oh and, Maxx, thanks. There are a lot of people who didn't know about KPOI who do now, because of you."

Claude Hall, whose column appeared every week in Record World magazine, wrote, *One of the best Top 40 radio stations in the United States is in Hawaii-KPOI. Sean Conrad is program director, and the station now includes Maxx Mahi Mahi, 5 to 9 a.m.; Jeff Kauffer, 9 to 1 p.m.; Sean Conrad, 1 to 3 p.m.; Glenn Martin, 3 to 6 p.m.; Riley Cardwell, 6 to 11 p.m.; and "Smackwater" Jack Waters from 11 p.m. to 5 a.m.*

We had a great-sounding radio station.

Just hanging out in my KPOI office, barefoot as usual.

Glen Martin

With Riley Cardwell, Spanky Lane (Rick Torcasso), Maxx Mahi Mahi, and Kirk Frosdick on the roof

Hawaii Summer Jam on Top 30 survey

With Cecilio

With Kenny Loggins and Jim Messina

Robyn and Rhonda with Kenny
Loggins and Jim Messina in
Hawaii in 1974

What's inside KPOI's Little
Grass Shack?

Vote KPOI entries

Honolulu Pulse, September 74

KPOI Business Card

Poimobile

KPOI

Maxx's prank newspaper clippings

Maxx clowning around at the Ala Wai Canal across the street from KPOI

CHAPTER 14

"DON'T LET THE SUN GO DOWN ON ME"

After a year of beating my head against the wall, trying to get the ratings up, I realized it was time to go back to the mainland and lick my wounds. So I quit, and Fred fired me. It was over. I was told that KPOI filed for bankruptcy after I left.

Our future was bleak. No job. No place to live. And since I wasn't making enough to live on in Hawaii, we were broke when we returned to LA.

Fred agreed to cover some of our moving expenses and airfare, but we wound up leaving most of our belongings in storage. After we left, my good friend, DJ Riley Cardwell, went to the storage unit, salvaged several boxes of personal belongings, and shipped them to us. Any valuable rock and roll memorabilia I had accumulated prior to that was now gone forever.

One doesn't just snap his fingers and get a radio job in LA. It takes time to get hired. We didn't have time, so I went to Harry Miller, who by then was Eric Chase, and PD of Drake's station, K100.

"Hey, Chase, have you heard of any openings around town?" I asked. "I need a gig."

"I have the 2 to 6 a.m. shift open if you're willing to do overnights. It doesn't pay much. Only $6 an hour."

"I'll take it," I replied.

I was a true un-success story. In a little over a year, I had gone from owning a house, a Porsche, attending parties with Hollywood stars, and being PD of KHJ to all-nights in LA. Needless to say, I was depressed.

Then came the issue of finding a place to live. The problem was solved temporarily by my old partner-in-crime, Don Branker. He owned a house in Playa Del Rey, which was right next to LAX. It was soon to be bulldozed to make room for a 747 landing strip. He and all his neighbors profited nicely when their houses were purchased by the city. He had moved to a beautiful new home in the San Fernando Valley and his old house was going to just sit there vacant for a month.

"Sean, I know you're hurting right now," Don said. "I don't mean to insult you, but you're welcome to live in the old house for a month if that would help."

The city had turned off the electricity to the neighborhood, but other than that minor inconvenience, we figured it would be tolerable and allow us some breathing room. A few sticks of furniture still remained, and we had running water and heat. I was hoping my girls were so young they wouldn't feel the pain we did. When I was young, and my family was poor, I never realized it.

After starting the 2 to 6 a.m. shift, for the first week, I stayed awake for seven days. I just could not sleep. Depression had a lot to do with it. K100 was located about fifteen floors up in a building at the corner of Hollywood and Vine. It was cater-corner to the Motown building and a few blocks away from Martoni's Restaurant, an Italian eating and drinking establishment frequented by all us Drake guys a few years earlier. Frank Sinatra went there often. The bar's back room was separated from the main dining room by a glass window. That back room was for the elite. The waiter who came with it looked like Robert Q. Lewis and had the quick-witted personality of Don Rickles. He was a put-down artist, and being the object of his scorn was a compliment. But then, I couldn't afford a Seven Up at Martoni's. To pick up some extra money, I took a noon to 6 p.m. Sunday afternoon shift at KGFJ, the rhythm and blues station on Wilshire, right across from the L.A. County Art Museum. So I could fit in with the mostly black audience, the PD gave me the name of David Washington. I enjoyed that gig because I loved the music.

After a month of this, Beth and I finally had enough money to rent a dumpy two-bedroom house, just around the corner from Normandie and Western Avenue in a questionable neighborhood. We had no furniture, but had a roof over our heads.

From the moment I got back to LA, I was on the lookout for another PD gig anywhere in California. By keeping my ear to the rail, I discovered that KSFX-FM in San Francisco was changing format from rock to a heavy emphasis on R& B and the newest form of music, disco. They were looking for a new PD.

KSFX was in the ABC-FM group as was my Chicago station, WDAI-FM. Group president Allen Shaw had always liked me and had respect for my work ethic. He had tried to get me to stay at WDAI but understood I couldn't turn down an offer like KHJ. I phoned him now and explained how I had made a huge mistake going to work for Paul Drew. I went into the details of KPOI and told him I was back on the mainland and wanted to apply for the KSFX job.

"If it were totally up to me, Sean, I'd hire you back in a minute," he said. "Let me call the new general manager, Don Platt, and tell him about you and your situation. Check back with me in a few days."

Before I had a chance to get back to Allen, I got a call from Don Platt.

"Allen Shaw thinks I should interview you for the PD opening," he said. "Can you catch a flight to San Francisco in the next day or so? We'll pay for the flight up."

Thank God for that. My alternative would have been hitch-hiking, and I probably would have done it.

I got the job. And hummed, "San Francisco, Here I Come," all the way back to LA.

After twelve years of an on-again-off-again marriage, Beth and I finally admitted it was over. We agreed it would be foolish for us to go to San Francisco together just to break up again. She had made new friends in LA, and we decided it was time to get a divorce. We had gotten married too young, and now, tough as it was, we both looked forward to a new life apart.

I felt terrible leaving my daughters behind but assured them as soon as I got an apartment in the city, I'd fly them up regularly. I left the car with Beth and started sending her child support right away. With a heavy heart, I said goodbye to my girls, and with nothing but a suitcase, moved everything I owned to the "City by the Bay."

My San Francisco E-Ticket Ride was about to begin.

CHAPTER 15

"PICK UP THE PIECES"

In 1975, KSFX-FM was located on the second floor of a nondescript two-story building at the corner of Polk and Stutter in an area that one hundred years before was called *Polk Gulch*. Polk Street, which paralleled Van Ness Blvd, started at Market and ended at Ghirardelli Square. It was a beehive of modern day restaurants and boutiques mixed in with leftovers from what was, at the turn of the century, a sea-faring community. On the first floor of our building, just under my window and a few yards down, was a drinking establishment called the Ship's Bell.

Walking in to that funky old bar was like walking in to another decade. It was a time machine. If you ordered a glass of wine, after the crusty bartender gave you a *What, are you kidding me?* glare, he'd blow the dust off a gallon jug of cheap wine and then reluctantly pour it in to a tall water glass. The Ship's Bell was where you came to drink whiskey and beer.

The only food available was a huge jar of pickled sausages sitting on the counter and small packages of beer nuts on the shelf behind the bartender. There was an antique beat-up pool table but no juke box. The contrast between this relic and what was going on just outside the door was astounding.

Polk Street was fast becoming the center of the San Francisco gay community. As I looked out my second-story window from my office, I saw colorful, noisy and festive people everywhere. It was a circus. From my vantage point, I saw things that could have been right out of a movie.

One warm summer afternoon I heard a commotion coming from the street and when I rolled my chair away from my desk to the window, I saw a four hundred pound man wearing nothing but a very large diaper, with a live boa constrictor wrapped around his rotund body. He was being taken for a walk by a skinny little guy holding on to what was apparently a collar and leash.

Another time, I heard tires screeching and saw a car speeding down an alley toward our building. He crossed over Polk Street and crashed in to a parked car just below my window. The car rolled onto its side up on the sidewalk. The driver then slammed his car into reverse and plowed in to a car parked on the opposite side of the street. He paused a few seconds and threw his car into drive and proceeded to crash into another one directly under my window, next to the first one he hit. It was worse than me in the KYNO Lincoln.

After a few more seconds, a bystander opened the passenger door and pulled this totally drunk maniac out of the car to prevent him from doing it again.

❧❧❧

No station had yet used disco music as the nucleus of a format. We carefully avoided saying the word "disco" on the air, knowing the side effects of boxing ourselves in like that could backfire later. We rotated new artists like Donna Summer and were first in the nation to play the long version of "Love to Love You Baby" in heavy rotation. As thanks for helping to launch her career, Casablanca Records gave me a gold record. Other core artists included The Ohio Players, The Village People, Kool and the Gang, K.C. and the Sunshine Band, The Average White Band (AWB), and Barry White.

At the same time, the artists who were already stars jumped on the disco band wagon and started to legitimize the music. Established artists like Diana Ross, Marvin Gaye, Blondie, and The Bee Gees churned out disco music left and right, which was perfect timing for what we were trying to do. Movies like *Saturday Night Fever* were a real shot in the arm. Even though KFRC and other Top 40 stations were mixing mainstream disco in to their format, we could take it to another dimension, which made us different, topical, and unique. Our ratings began to climb, and I was once again on a roll.

Shortly after arriving at KSFX, I was visited by promo man Henry Coleman from Liberty Records. Henry and I became great friends. He was a native San Franciscan and knew all the nooks and crannies of that fabulous city. He helped me get my first apartment on Baker Street, between Sacramento and California streets. And he knew the best hole-in-the-wall Italian delis, where you could get a huge Italian cold cut submarine sandwich for only a dollar-fifty.

Henry and I wound up in Sausalito one day while he was showing me around. We stopped into Paterson's Bar on Bridgeway for a drink. Paterson's is a funky old Scottish Pub that has been in downtown Sausalito for many years. Its tartan pattern wall-to-wall carpet and row upon row of dusty scotch bottles reek of the '60s.

We walked in on a Saturday afternoon to an empty bar except for one old guy sitting by himself at a table by the front window. He was totally unkempt, dressed like a bum, with long gray hair and a beard.

Henry was an outgoing, friendly guy who would strike up a conversation with a stranger at the drop of a hat. As we took a seat at a table next to this curmudgeon, Henry asked what he was drinking.

"Thirty year-old scotch," he growled back at us.

"I'm Henry Coleman. This is my friend Sean. You look familiar."

"Yeah, so what," he muttered as he turned away and hunched down over his drink.

"Wait a minute," Henry said. "You're Sterling Hayden. I've seen you in so many movies. What are you doing here?"

"I live here." He pointed out toward the water. "On a boat."

After several hours of pounding down scotch and deep in conversation with Hayden, we discovered his passion was sailing, and being in the movies was just a means to an end. He professed distaste for film acting and was unimpressed with being a movie star. He was an accomplished author and a former marine. His movie credits included the *Asphalt Jungle, Dr. Strangelove*, and *the Godfather*, where he played the corrupt cop that Al Pacino wasted in the Italian Restaurant.

Just a few blocks down Polk Street was the La Piazza Italian Restaurant owned by an Iranian named Mike Deeb. Mike and I became great friends. One beautiful sunny San Francisco day in 1976, I was having lunch with some station people when I noticed a striking blonde having lunch with friends at another table. I was not the most aggressive guy when it came to pursuing women but something about her registered. About a week later, it was the same scenario. This time, I called my buddy Mike over.

"Hey Mike, what do you know about that sweet young *thang* sitting over there?"

"Well, I don't know her name, but I think she works a few blocks from here at the county building," he said.

"Would you mind sending her a glass of wine, put it on my bill, and tell her it is from me?"

"Sure, no problem."

This was back in the day when it was acceptable to have a glass of wine or two at lunch, then return to work without getting fired. Eager to see her response, I watched as Mike sat it down in front of this angel. He bent over, said something to her, and then pointed over at me. She picked up the glass and, with a smile on her face, toasted me. Touchdown. When I was ready to leave, I stopped by her table.

"Hi," I said. "My name is Sean Conrad."

"I'm Birgitta Hallgren. Nice to meet you."

"I've seen you here before. I work a block away so I'm here for lunch all the time. I love this restaurant, and Mike, the owner, is a great guy."

"I work in the county building on Van Ness," she said. "I'm not here as often since I have to drive to get here, and then I have to deal with the parking issue. Where do you work?"

"I'm the program director of KSFX-FM Radio a few blocks away. You know, I'd love to see you sometime. Can I get your phone number?"

I was ecstatic when she gave it to me.

Birgitta Ingrid Hallgren was only twenty-two years old and from good Scandinavian stock. Her parents immigrated to America when she was three. They raised her with great moral values and a strong work ethic. She was beautiful inside and out, and I fell for her instantly. A few months later, we rented a house in Corte Madera, five miles north of the Golden Gate Bridge. I told Birgitta that I wanted my girls to move in with us. When I saw them last, they were starting to show signs of becoming street hardened, and I couldn't let them live in Southern California any longer.

"I love those girls," she said. "I agree. They need to live with us."

"Then I'll call Beth and see how she feels about it. I really think she will be okay with it."

Beth was involved with someone, but other than my child support, as far as I knew, she had no money coming in. She agreed that our daughters would be much better off living with us in Marin County than living on the poor side of town in LA.

Birgitta was the best thing that ever happened to the girls and the perfect antidote for what they were being exposed to. We were one happy little family with a bulldog puppy named Shaka, living in Marin County.

I bought a new 1976 black, metal flake Toyota Celica that looked a lot like a Porsche, and was a real eye catcher. Birgitta and I commuted across the Golden Gate Bridge to work every day and I was back on top of the world again.

Soon we purchased a three-bedroom house with a pool in a community called Mariner Cove, in Corte Madera. Not long after that, we had a huge wedding at the Sweden-Borgen Church in San Francisco with a reception in Sausalito.

I invited about one hundred of my wildest and craziest friends from the radio and record industries and I chose two best men. One was Russ Terrana. The other was Ralph Tashjain, who was a native San Franciscan and a very funny guy. Ralph's family owned Tashjain Florists at the corner of 21st and Mission. He had gravitated toward the music business and was a promotion man who hammered me for air play regularly.

ᗉᎧᗉᎧ

One afternoon at KSFX, Darla Jensen, the receptionist, called me on the intercom. "Sean, Francis Ford Coppola is here to see you."

"Francis Ford Coppola? Are you kidding me? Why? What for?"

"He says he wants to talk to you about an idea he has."

In 1972, I was first in line to see *The Godfather* in Fresno. I had read the book and completely connected with the authentic, first generation, right-off-the-boat, Italian-ness of the movie. My grandmother was pregnant with my mother when she came to Ellis Island from Italy.

During my childhood, on the weekends when I wasn't with my father, I was at my Italian grandparents' house. From those experiences, it was obvious that *The Godfather* movie was birthed by true Italians, not the Hollywood version. And when Francis paid me his little visit, it had only been two years since he had made *Godfather Two*. The fact that he was in the lobby, asking for me, was inconceivable.

"I'll be right there." I bolted down the hall out of my office, turned the corner, and there he was. The Godfather himself.

"Hi, F–Francis, it is very nice to m—meet you, sir," I stammered. "Come on down to my office."

He followed me the short distance, immediately went to my desk chair, and sat down. That seemed to be an invitation for me to sit in the chair where, normally, a visitor would have been seated. Here I was, facing my desk with Francis in my chair looking at me. I had a dart board in my office that was attached to the wall directly above the seat I was sitting in, so that when I was bored, I could toss darts from behind my desk. Francis started throwing darts.

"Sean, I need to pick your brain for a project I'm considering. When I was a kid, I used to listen to radio when all the programming was originated live from the studio. None of this pre-recorded shit like they do today. I'm thinking of buying a San Francisco radio station, possibly KMPC, and going all live with it. Live music, live drama, live commercials, all live programming. What do you think?"

"Well, Francis, I think that would be a pretty difficult thing to do," I said. "Consider the cost of having a full time orchestra in the studio, a full staff of voice actors, and the union issues you'd have to deal with. I can't imagine how you could ever make any money doing that."

We then went through all the reasons why it would be financial suicide and logistically impossible to do. After discussing the project and a tour of the station, he left, never to be heard from again. To the best of my knowledge, he dropped the idea that day and moved on to his next project: some silly little theatrical production called *Apocalypse Now*.

In August of that same year, Elvis died, and so did my friend and mentor, Bob Holliday, who was involved in a fatal car accident in Tucson. Bob had called me just to chat only a few months before. While *he* never made it past small to medium radio markets in his career, *I* was now programming a radio station in my third major market city, and none of that would have ever happened if it weren't for Bob Holliday.

പ്രാ

In 1976, Stevie Wonder was about as hot as an artist could be. In 1972, 1973, and 1974, he had two gold and one platinum LPs. By 1976, two years had gone by since the public had heard any new material from him. Meanwhile, at the Motown studios, Stevie was busy recording what would become his first double platinum LP, a two-disc album called *Songs in the Key of Life*. Back in those days, radio stations prided themselves in getting copies of new records by super-hot artists before the competing radio stations could get copies to play.

There were various ways to do this, like telling the record company you'd add one of their stiffs, a record that had no chance of being a hit, if they would slip you a new record by a hot artist a few days before giving it to the competition. Then you could play it on the air while bragging about the fact that no other station in town had it, therefore rubbing the competition's noses in it. It was a game we programmers played, and it was fun.

I had gotten a call from Eric Chase, who was Harry Miller, is now Paul Christy, and was the PD of KGB, in San Diego. Eric said he had a reel-to-reel tape of side one of *Songs in the Key of Life*. Apparently it was a rough mix that was tossed into a trash can after a session but was smuggled out of Motown Studios by someone inside. Eric said he wouldn't touch it with a ten-foot pole because he didn't want a lawsuit, but that if I wanted it, he'd ship it up to me.

I talked to my general manager, Don Platt, who agreed to take a chance and let me play it. The next day, it was in heavy rotation, and we were the only station, not only in San Francisco, but in the world, to be playing it. Motown lawyers made some feeble attempt to stop us but chose instead to just rush the finished product out to the radio stations and the stores. We got away with it, and Stevie Wonder's *Songs in the Key of Life* was a big hit with our listeners.

FRANCIS!

With Birgitta, Mom, and Ralph
Tashjain at wedding number 2

Marc Nathan (Casablanca Records)
and friend with my music director,
Coleen. Marc was promoting the
latest Buddy Miles LP.

Partying down with Birgitta,
Gary Cocker,(Helmet Kerling)
and Mary Kilmartin

The poster says it all
about the format.

My Donna Summer "Love to Love
You Baby" Gold Record

CHAPTER 16

"PLAY THAT FUNKY MUSIC, WHITE BOY"

On Tuesday, May 17, 1977, an estimated two thousand National Association of Broadcast Employees and Technicians Union (NABET) members went on strike at all ABC-owned-and-operated radio stations.

At a non-union radio station, the jock on the air did two things at once. He was talking and concentrating on what to say next, and at the same time, he was operating all the equipment. It's like being a singer and, at the same time, playing the guitar. But at an NABET radio station, the jock on the air was strictly forbidden to even touch the equipment. He was allowed only to point to a NABET engineer to start a record or a commercial. When he needed the microphone turned on, he gestured with a hand.

Some liked it. I didn't. Different engineers had different temperaments. Some were an absolute joy to work with. The

ones who were assholes were the most difficult because com-
plaining about an uncooperative engineer got you nowhere.

It only made matters worse since they would then start
missing cues, reacting slower than normal, and on occasion,
actually falling asleep. I had that happen. If for some reason, a
jock did break a union rule, like hit "start" on a cart machine
because the engineer left the room to go to the bathroom and
the record had ended, he'd be slapped with a grievance. I
avoided that at all costs. Management could not fire an engineer without building a case for at least a year.

They didn't care about ratings because whether those ratings were sky high or in the toilet, their jobs were secure. One
of our engineers had been in the union so long that he would
work six months, followed by six months of vacation with full
pay. I always had to walk on eggshells with them and kiss their
asses if I wanted them to work with me.

All radio stations have a music room where the 45s and
LPs are cataloged and stored. In a NABET shop, non-union
personnel are strictly forbidden from touching the records in
that room. If we needed an LP, we had to ask the music librarian to hand it to us. The music librarian was an official NA-
BET position. We were not allowed to walk past her desk.

It was like a weird version of how many people it takes to
change a light bulb. When a DJ was assigned a commercial to
record, it took three people to do it. One to operate the equipment, one to voice the commercial, and a so-called producer to
stand over them with a stopwatch and say, "Okay, go." In a
non-union radio station, this whole process was performed by
one person.

When the engineers walked because they wanted better
benefits, less work, more vacations, and more money, ABC
locked them out. Secretaries, sales people, and management
were all trained on how to operate the equipment. For four

months, we had to cross their picket lines in order to get to work. Many of us worked twelve- to sixteen- hour days, seven days a week. Not only did the station sound better than ever, the ratings went up, and we had a ball.

We all came together as a team and worked our asses off. ABC wanted us to be happy, so every day for lunch, a delivery guy from David's Deli, an authentic New York-style Jewish restaurant a few blocks away, delivered a huge basket of gourmet sandwiches, soups, and desserts. For those of us working after regular business hours, ABC had a charge account at Dario's Pizza, and we could order delivery of anything on their extensive menu any time of day. They kept the refrigerator in the lunchroom stocked with sodas, beer, and wine. They also doubled everyone's pay. I went from $600 to $1,200 a week, and that went on for four months.

Of course, if we needed a little assistance staying awake all those hours, the local coke dealer was only a phone call away. Our chief engineer, Steve Roberts, always had his door closed. But if we knew the secret knock, he'd yell, "Come in," and pull the giant mirror out from under his desk to carve up a line or two. We were all overloaded and stressed to the max, but at least we were having fun.

It didn't last forever. The union and the cumbersome restrictions returned. Well fed, well paid, and ready for a rest, I was able once more to focus on the programming and the music.

As usual, I was pursued by record promoters. One of my good friends was promo man Joel Newman, who was promoting the disco tune "Play That Funky Music White Boy" by Wild Cherry.

"Joel Newman is here to see you," Darla said over the intercom. "He says he's got someone he wants you to meet."

"Send him in."

Joel was not alone. Beside him was a woman wearing a long black trench coat and his photographer Pat Johnson. Joel shut my door.

"Hey, what's going on?" I demanded.

"Shut up," he said. "Come over here and sit down. Don't say a word."

I got out from behind my desk and went over to the seat Francis Ford had me sit in. Joel then walked over behind my desk as Francis Ford had done, but instead of throwing my darts, he put a 45 RPM record on my turntable and started it up.

With the volume cranked to the max, "Play That Funky Music White Boy" started thumping through my speakers while the woman in the trench coat dropped it to the floor. Totally naked, she started bumping and grinding to the funky beat right there before my eyes.

I have pictures to prove it. Pat Johnson took them.

Startled I jumped up from my seat.

"Go sit in your chair," Joel said, the music still blaring.

"Whatever you say," I told him.

The stripper made her way to me, and I experienced my first and only lap dance. Okay, I did add the record. I would have anyway because it became a monster hit.

ΩΘΩ

Within the first six months of my job as PD, I succumbed to the peer pressure all around me at KSFX to attend EST, Erhard Seminar Training. It was marketed as a large group awareness training program. Over the course of two weekends, paying participants were exposed to a cross between Zen and scientology. During the grueling ten-hour sessions with hundreds of other people squished together on hard metal chairs, the EST trainer would dish out abusive, demeaning, gibberish that was meant to cause us to *get it*. We were not permitted to go to the bathroom and had to go all day without food.

The first EST training took place before one thousand people at San Francisco's Jack Tar Hotel in 1971, then it spread all over the world. My training took place there, too, since the building was located only two blocks from KSFX. At the end of the three days, we were instructed to sprawl out on the floor, close our eyes, kick, scream, and *let go* for about an hour. Some 700,000 people took the training between 1971 and 1984.

Back at the station, since more than half the staff had taken the training, a divide developed between those of us who *got it* and those who didn't. If there was a conflict or disagreement between two people who were trained and two who were not, the two EST-holes could look at each other and wink, as if they knew something the other two didn't. I bought into this brainwashing for a few months but soon felt like it had been a colossal waste of time and money.

Play That Funky Music, White Boy!

With General Manager, Don
Platt, and Marty Feldman

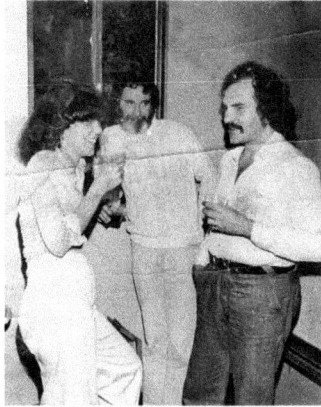

With mid-day jock, Barbara Harrison
and Chief Engineer, Steve Roberts
Barbara is now the morning news
anchor for the NBC 4 TV affiliate
in Washington D.C.

CHAPTER 17

"DISCO LADY"

While programming KSFX from 1975 through 1978, I worked with some gifted people: Tommy Saunders, mornings; Barbara Harrison, mid-days; Ron Samuels, afternoons; and Eileen Fields in the evenings. Our music librarian, Mary Kilmartin, was the best, and I was proud to hire talented women as more than clerical support. Equal opportunity might be difficult to carry out on government forms, but at my station, it worked the way it was supposed to, not because of me, but because of them.

About six months after I was hired, the music director I inherited from the previous regime moved on to another station, leaving the position open. Enter Darla. I was sitting in my office smoking a cigarette, as most everyone did back then, feeling the pressure all major market PDs experienced every day, when Darla walked in. She took a seat on one of my visitor chairs, seductively crossed her legs and, as she chewed gum, said, "I'm on my lunch break right now, and I want to apply for the music director opening. What do I need to do?"

"Uh-huh. Is that front desk crap getting to you?" I asked.

"I hate it, and I've paid my dues, and I love music, and I'm perfectly qualified to fill this position, and I—"

"Hold on, Darla," I said. "Slow down. I know you're a music junkie, and I'm sure you'd do a great job, but you know what I have to do before I can hire someone. You've taken the calls and seen the people flock to my office anytime I have an opening. Where do you think they come from?"

"An ad in the paper?"

"Yes, that too, but mainly, ABC insists the job be posted throughout the corporation and all the broadcast publications. It's an equal opportunity thing. I suggest you fill out an application and stand in line. If it were just my decision, I'd rather hire you than someone I don't know, but I can't."

"But in the end, it's you who makes the decision, and I hope you hire me."

I handed her an application.

"I have to play the corporate game, so fill this out and I'll put it in the stack," I told her. "According to company rules, the position must be left unfilled for a 30-day period so everybody and their parrot can have a chance to apply. Just keep doing what you're doing at the front desk and let me work through this."

True, I was attracted to Darla. I also felt she would make a great music director. She was young, she knew the music, and loved to go to concerts and new release parties the record companies were always throwing. After a month of interviewing qualified and mostly unqualified people, I was able to justify her as the best person for the job, because she knew KSFX inside and out and had a handle on our audience. Darla got the job.

Since the KSFX format was background music for dancing and partying, we were willing participants in the nightlife
that came along with it. On many occasions, I had to attend
station-sponsored events held at a disco, concert, or record
company cocktail party to introduce a new artist. Although
Birgitta attended some of these with me, many times I went
with Darla.

As music director, a large part of her duties were to meet
with the record company reps who needed to pitch their product. A record promoter's job was to do whatever it takes to get
their records added to radio station playlists. Every Monday
and Tuesday, these guys would make their rounds from station
to station and from record magazine to record magazine. Dealing with record people and sifting through the hundreds of new
records the industry released every week was a time-
consuming task and one I didn't have time for.

Darla's job was to give each record company representative fifteen to twenty minutes face time in her office and allow
them to audition their new releases. Then on a weekly basis,
she would comb through the Billboard, Record World, Gavin
Report, and Radio & Records Top 100 Music Charts to see
what was hot and what was not. Every Wednesday afternoon,
after she had weeded out the obvious stiffs, we would sit in my
office and audition all the new music she thought might fit the
format and should be added to the KSFX playlist.

Some of the promotion guys didn't like the fact that she
kept me insulated from direct contact with them, but that was
her job. They knew she was not the decision maker and that,
ultimately, I was. The easiest way for them to get to me was to
invite us both to lunch—usually a very liquid lunch at the
some of the finest restaurants in town—and the bill was always paid by the record company.

Promo guys were a hoot. They were hired because of their persuasive personalities and were all characters in their own ways. They'd pack our noses, take us to restaurants, sporting events, or to wherever the lights were the brightest and the music the loudest. The guys who had hammered me for air play at KYNO started coming out of the woodwork. Although they visited me frequently in Fresno, which was part of their territory, their home base was San Francisco. On a weekly basis, my old friends Dino and Johnny Barbis, Jeff Trager, Kenny Reuther, Mike Kolodovich, Bruce Sperling, Bruce Shindler, Pete Marino, Joel Newman, and Larry Carp started showing up at KSFX to share Fresno stories and try to get airplay.

Dino Barbis later became vice president of Warner Brothers Records, but his biggest claim to fame was saying, "Ya-da-ya-da-ya-da" to make a point, in all his conversations. He had picked the phrase up from Lenny Bruce and was known for mouthing it regularly long before Seinfeld popularized it in the '90s.

ᏟᎦᏟᎦ

Because of all the nightlife and job-related partying with Darla, I found it difficult to resist physical involvement. My strong attraction to this free spirit was intense. She was only twenty-one, about five-foot-five, freckly, and with an explosion of curly strawberry-blonde hair. Although she had stated several times that a relationship with a co-worker was out of the question, that rule took a different path once I got married. I'm not sure if she was jealous, or if she just liked the drama of being with a married man.

One day, a few weeks before my wedding, Birgitta was sitting in my office during a lunch break when Darla charged in and sat on my lap, throwing her arms around my neck. Talk about uncomfortable. All I could assume was that it was some sort of woman-to-woman message that Darla was sending. Birgitta shrugged it off and never mentioned it.

In the years that I knew her, between 1975 and 1981, I don't think Darla ever really fell in love with anyone—certainly not with me. There were times it seemed that the nicer I treated her, the more distant she would become. I noticed that she was more attracted to the bad boys than to my type. Since it's human nature to want what you can't have, I had to have her, and she knew it.

One night at the Mandarin Restaurant, high atop Ghirardelli Square, overlooking the Bay in San Francisco, I again expressed my desire to be intimate with her.

"Okay, Sean," she said, "if you want me to be your mistress, I will. No big deal. You say when."

That night at the Fisherman's Wharf Hotel, right down the street from the Mandarin, I finally got my wish. I have no excuses for my behavior. At age thirty-one, I still had about fifteen more years before my brain relocated to my head.

Darla and I worked together at KSFX, staying under the radar with our secret affair, until late in 1977 when there was a shakeup at the station. Disco music and all its polyester leisure suits were passé, and the ratings were dropping because of it. Management decided it was time to change to an AOR format. I was fired, and Darla soon followed. One would have to be deaf, dumb, and blind not to realize she and I were fooling around, so that probably affected the management decision to let us both go.

Many times, up-and-coming recording artists would be brought up to the station by the promo guys to do a meet-and-greet with us. A few months before we were let go, Darla had made friends with the manager of Vicky Sue Robinson, whose hit, "Turn the Beat Around," was in heavy rotation. I remember hearing from promo men in the industry that this group was into serious drugs.

Darla followed the Vicky Sue Robinson crew to New York City and blended in with the nightclub scene. I stayed home, smoked dope, and watched *Family Feud*, and other brainless daytime TV shows. I was trying to heal, but I wasn't sure from what.

Backstage at the Great American Music Hall for the Tom Scott and The LA Express gig. Bottom right-hand Corner, my good friend Joel Newman, Epic Records

From the Pink Section of the San Francisco Chronicle

CHAPTER 18

"JUST WHAT I NEEDED"

I was at a crossroads in many ways. Darla was gone. I did not want to destroy my marriage and continue to hurt Birgitta. I also had decisions to make about my career. How much longer did I want to be a major-market PD? I had many friends in the record business and thought I'd make a good promo man. Joe Smith went from being an obscure Boston disc jockey to CEO of Elektra Asylum Records, then later CEO of Capitol Records.

In early 1978, my friend and local independent promotion man, Larry Karp, told me about an opening for a regional promotion man at Motown records. Motown? Hell, I was already connected to the Motown family from my days in Detroit. This would be a natural. I called Russ and was told he was in a session.

"Oh, wait," the receptionist said. "He just walked by." I could hear her saying to Russ, "Sean Conrad is calling for you. Can you take the call?"

"Yeah, send the call to the studio."

"Hey, paison," he said. "What's happening?"

"What are you doing?" I asked, unsure of how to begin.

"I'm in a session with Jermaine Jackson right now. He's recording a solo album. It's pretty damn good, too."

"Has it got a hit single?" I asked.

"I think so. It's a cut called "Let's Be Young Tonight." He's also re-doing the old Shep and the Limelight's hit, "Daddy's Home." I know that one's a hit."

"Hey, Russ," I said, trying to sound calm and cool. "I heard there's an opening for a Motown West Coast regional promo man based out of San Francisco."

"What about KSFX?" he asked.

"They decided to change format. I'm out."

"That's fucked up."

"Not really," I told him, feeling more comfortable with the truth. "I was burned out anyway. Besides, I'm looking for a change. The record industry interests me. Who better to go to radio stations and hype records?"

"Right on. I'm sure you could out-hype any of them." He laughed, and his confidence encouraged me.

"Who should I contact at Motown?" I asked.

"Call Barney Ales. You remember him from Detroit? I'll let him know you'll be in touch."

"Thanks, Russ," I said. "I owe you a big fat one.""

"You're on."

Barney Ales had been with Motown since the beginning, and he controlled the ebb and flow of the record product. I flew down there, went through the interview, and got the job. Motown was notoriously cheap, so the pay sucked, but my title was West Coast Regional Pop Promotion Manager, which was quite the feather in my cap. My territory was Seattle, Denver, San Francisco, and many other surrounding towns.

Before I flew back to San Francisco, I drove over to Motown's Romaine Street Studios and popped in on Russ, who was in a mixing session with Smokey Robinson. I entered the building, came up to the glass wall separating the lobby from the interior, and told the receptionist who I was and that I wanted to see Russ. I had been to this building many times before when I lived in LA. It contained separate recording studios and several small mixing rooms.

A few minutes later, Russ came to get me from the lobby, and we headed to Twilight, his favorite mixing studio upstairs.

"Well, did you get the job?" he asked.

"Yep. We both work for Motown now. If you ever told me back in Detroit that something like this would happen, I'd have called you crazy."

Just before we went in to the studio, I saw Jermaine Jackson walking down the hall wearing a white T-shirt with big block letters on the front that said nothing but FUCK YOU. Boy, did he get up on the wrong side of the bed.

In the mixing room, Smokey was standing over the console sliding some dials up and down, listening to his music.

"Smokey, this is Sean Conrad," Russ said. "You remember him from WKNR back in Detroit, don't you?"

"I sure do. I think we met at one of my recording sessions, didn't we?"

"Yeah, we did." I didn't think fast enough to tell him how I bought his first hit record with forty-nine cents and six Pepsi bottle caps.

"I remember when you use to play my new records before anyone else did," he said. "Have a seat, man."

I was being invited to sit in and watch as Russ and Smokey did the final mix on his new LP, *Smokin'*.

"Hey Russ," I said. "Here's that big fat doobie I owe you." We all lit up, and I watched and listened as Smokey's sweet, sweet music came together. After the session was over, Russ took me aside and asked me when I had to go back to the city.

"Well, my flight leaves in three hours," I said.

"Stick around. Fly back tomorrow. Hal Davis and I are going out club hopping in his big limo tonight. Come with us."

I hadn't seen Hal since the party at the Jackson Five's home back in 1974. He came from a poor Detroit neighborhood and walked with a slight limp from a bout with childhood polio. Women were hanging on him all the time. He wrote and produced "I'll Be There" and "Dancing Machine" for the Jackson Five, and many other songs that the Supremes, The Temptations, Gladys Knight, Little Stevie, Marvin Gaye, and the Four Tops recorded.

"What the fuck? Why not?" My decision made itself. "I'll go home tomorrow."

Later that evening, Hal pulled up in his stretch limo, and Russ and I piled in.

"My man. What's happenin', Sean?" Hal said as he gave me the official fist wrap hand shake. "I hear you're workin' for Motown now. That's cool, man. Let's go do some celebratin'."

We went club hopping from one all-black nightclub to another. Between destinations, we'd snort a few lines and raid the limo wet bar. Hal was on a talent hunt, which was the purpose of all this debauchery.

Working for Motown was fine except for the pay. On the company credit card, I was constantly flying to Seattle and Denver and driving to California valley towns like Fresno and Stockton. Some of the records I was in charge of getting played were "Three Times a Lady" by the Commodores, "You and I" by Rick James, "Rock With You" by Michael Jackson, and a Diana Ross record called "The Boss," that mid-charted.

One of the highlights of being with Motown was the time I went to a company convention held at Caesars Palace in Las Vegas. By night, Russ and I wined and dined, and hung out with Diana Ross and Berry Gordy. By day, we attended meetings on how best to promote Motown Records.

e⁊e⁊

I had been with Motown for close to six months when I heard about another job opening from Les Garland, who was the PD of KFRC in San Francisco. During one of my weekly visits to see Les and his music director, Dave Sholin, Les told me a position had just opened up at Elektra Asylum Records. Les was good friends with Sammy Alfano at Elektra Asylum, which pretty much got me the job. The pay was nearly double, and the roster of recording artists included the Eagles, Linda Ronstadt, Queen, Joe Cocker, Tom Waits, Ronny Montrose, and Jackson Browne.

The first thing Kenny Batiste, head of A & R at Elektra, did when he took me in to his office for the interview in LA was to play a record by a new group they had just signed and recorded out of Boston called the Cars. "Just What I Needed" was the title and I knew it was a hit.

"So, what do you think?" Kenny asked. "Is that a winner or what?"

"Damn, Kenny, I love it. I can tell you with my program director ear, it will be number one."

Evidently, that was what he wanted to hear because he hired me. He then gave me a tour of the offices. We ran into Don Felder and Glen Frey of the Eagles, who welcomed me aboard, on the way to the human resources department, where I filled out the paperwork before heading to the airport.

When I got home to Corte Madera, I was exhausted, but not so tired I couldn't take Birgitta and the girls out to a steak dinner on my new expense account in San Rafael. We talked about how this job was superior to the Motown job, especially since I wouldn't be traveling all the time.

"Dad, do you think we'll ever get to see Queen?" Rhonda asked, her green eyes bright.

"Yeah, Dad." Robyn chirped. "I love Freddie Mercury."

"I don't know about meeting them, but you'll get to see them because they're coming to town in December."

"Wow, Dad, and we get to go?" Robyn asked.

"I wouldn't leave you at home for this show. Yep, I'll figure out a way to bring all three of you."

Finally, my marriage was working. My family was working. I was going to save both with this job, and maybe even myself.

The San Francisco Elektra Asylum offices were located at Hyde and Jefferson, catty-corner to the Buena Vista Bar & Grill, where the first Irish coffee was poured over a hundred years before. It was across the street from the Hyde Street hill cable car turn table. Our offices were in the historic Cannery Building, which was built in 1907 and had been retro-fitted to contain our newer structures. We were a few hundred yards from Ghirardelli Square and about the same distance from where the street vendors sold fresh crab and shrimp. At lunch, I'd head to the rear of the building, take a 100-year-old creaky freight elevator down to ground level, and walk a short distance to some great Italian seafood restaurants, including Di-Maggios, Aliotos, and Scomas.

My boss, Bill Perasso, was a native third-generation San Franciscan with a deep Italian heritage. He looked like a character out of a Clint Eastwood spaghetti western. Twenty of us worked in the WEA office, and every Monday at 5 p.m., we'd have a catered all-staff meeting in his office to plot strategy on getting new records on the air. It was all paid for by the three record companies we represented.

One week, Bruce Hicks from Warner Brothers was responsible for buying the food. The next week, it was Steve Feldman from Atlantic Records. Then, I would be the guy hosting the meeting, paid for by Elektra Records. Since there were no limitations on how much we could spend, we'd try to top each other with extraordinary offerings each week.

When one of my artists came to town for a concert, the pace quickened. The Queen concert was scheduled at the peak of Queen's career. The *Fat Bottomed Girls* LP had just come out, and almost everyone wanted to see them live. The concert sold out right away. About a week before the concert, Ron Lanham, Elektra Asylum West Coast promotion manager, called me from LA to explain what I needed to do while the guys were in town.

"Okay, Sean," he said. "Brian May, Queen lead guitarist and songwriter, is the only member who'll come to town the day before the concert."

"What time does he arrive, and at which terminal?" I asked.

"His flight gets in from London at 9 a.m., and you need to pick him up at the International Terminal, the last building as you make the big circle."

"Any idea how much luggage he'll have?" I could feel myself get nervous. "My Celica doesn't have a lot of trunk space."

"Get a limo. Keep it for a couple of days. Use the company credit card. After you get him checked in at the Marriott, get right over to KMEL and KSAN. They're going to interview him live on the air."

"Cool, but we'll probably get done with all that pretty early. Then what?"

"You'll have a limo," he said. "Brian loves North Beach. Take him to lunch at Fior de' Italia, then go wherever he wants to go."

"I can handle that."

Brian turned out to be the nicest guy. No bloated ego. Just a regular person.

"After we're done with the radio stations, let's hit the antique shops on Grant Street in North Beach," he said in his proper British accent.

"We got all day," I told him. "You say where, and I'll get you there."

After he had spent a few thousand dollars on antiques for his home in England, I dropped him off at the hotel and headed home in the limo for some rest before the concert.

Concert day finally came on December 16, 1978 at the Oakland Coliseum. As promised, we took Rhonda and Robyn with us in the limo. The year before this concert, "We Will Rock You" had been released, and that's what they did that night. They brought the house down. The sound level was ear deafening, and Freddie was as astounding as always.

After the show, I had made reservations at the Empress of China Restaurant in the heart of San Francisco's Chinatown. Birgitta and I, along with Rhonda and Robyn, took the limo to the restaurant. Since it was going to be a very late night, we had the limo take the girls home, and then return to get us. The Empress of China was known as the finest Chinese restaurant in San Francisco at the time. At the long dinner table in a private room, twelve people including band members Freddie Mercury, bassist John Deacon, and Elektra personnel enjoyed an exotic meal to the tune of over a grand, which was a lot of money in 1979.

On August 19th of that year I had to do a meet-and-greet with listeners at a Hank Williams, Jr. concert. It took place at the Marin Civic Center when Hank was red hot with his hit song, "Family Tradition." As a kid, I grew up in a home where we woke every morning to the sounds of Hank Williams, Sr. coming out of the small table top record player my dad owned. I was not a Hank, Sr. fan. I was an Elvis fan. Hank, Sr. and Elvis may have both hailed from the deep south, but in the late 'fifties, their music was as different as black and white, day and night.

At the concert, after the listeners had departed, only Hank, Jr. and I were left in the room to wait for his turn to go on stage. As we passed a bottle of Jim Beam, sippin' whiskey back and forth, the same whiskey he sang about in so many of his songs, I couldn't help but reflect on the generational differences we both shared with our fathers.

"Hank, I got to tell you, man," I said, "I am blown away to be sitting in the same room with the son of my father's favorite country music artist."

With his feet propped up on the table, leaning back in his chair, he said, "Is that right? Well, Dad had a lot of devoted fans."

Hank, Jr. was a wild man back then, surviving a near fatal fall off the side of Ajax Mountain in Montana four years earlier. Had he not survived that fall, he would have died at the age of thirty, the same age that his father had, not from a heroin overdose but a freak accident. "My dad was born in Alabama," he said. "Where'd your father come from?"

"Crossville, Tennessee," I replied.

"A couple of good ole' country boys. Well, time to go knock 'em dead. Good meeting you, Sean."

Hank, Jr. yanked that audience of around fifteen hundred fans onto their feet, stomping and clapping, and just getting wild. Five years later, he hit it big again with "All My Rowdy Friends Are Comin' Over Tonight," a song that was the ABC Monday Night Football opener for more than twenty-five years.

⁊⁊⁊

I had a special place in my heart for the Cars, since they were the first new Elektra Asylum group I promoted. Bostonians Ric Ocasek, Benjamin Orr, David Benjamin, Elliot Easton, and Greg Hawkes loved to come to San Francisco. When they were in town, I'd meet them at the airport, usher them to their hotels, and drag them around from one radio station to another. They especially loved being taken to the Hunan Restaurant in North Beach when it was still a hole in the wall.

The night they were scheduled to appear in Fresno, we all holed up at the airport Hilton before departure the next day. We had about six rooms and turned on the party lights that night. Many mostly female fans showed up.

"Hey, Sean, want to hear some of my new songs?" Ric asked me at one point in the evening.

"I'd love to. Where?"

"Follow me." He led me to another room, where a cassette boom box sat on a table, and he played me some raw, unproduced basement recordings of music that would eventually be on their second LP, *Candy-O*. It was released June 2, 1979.

⁊⁊⁊

About this time, the annual Warner-Elektra-Atlantic music convention was held in LA. The sales and promotion teams from all over the country were flown in to be exposed to the new releases that were planned for the next twelve-month period. Music videos were in their infancy back then, but we were shown a video of Fleetwood Mac's *Tusk* LP, which was ahead of the times. The Eagles' new LP, *The Long Run,* was presented to us along with many other albums by Devo, Tom Waits, Eddie Rabbit, David Gates, and Bruce Springsteen.

A few days before we left for LA, Darla called, just to say hi. Darla never called just to say hi. I was curious and more intrigued than I should have been.

"So, what have you been up to, Sean?"

"Just working. We're all flying to LA Friday for the annual convention. What about you? Are you still in New York?"

"No. I got burned out on that whole thing. I just had to get away, so I moved to San Diego. I tried to get a job in radio, but there were no openings. I got a secretarial job, and I'm living with friends on Balboa Island."

"That's a pretty expensive place to live," I said. "You must be making good money."

"Not really. There are four of us in the house, and we split the rent four ways. Where are you staying while you're in LA?"

"Century Plaza. That's where the convention is being held. too." I took a deep breath. "Why don't you drive up Sunday night? I'll buy you dinner. You still drive that little B-210?"

"Sure do. I love my little hatchback. I'll see you there," she said. "But, Sean, let's not go to dinner. I'll bring a six-pack, and we can just hang around the room."

Although I knew I should have said, "No," the answer I gave her was, "Why not?"

Delivering the new Eagles record
"The Long Run" to KLIV in
San Jose, California

Bill Perasso

Left to right, Don Wright, Tawn
Mastery, and Paul (Lobster)
Wells at KSJO receiving a Cars
gold record in 1979

With The Cars during a live
KSAN interview

Freddie Mercury on Stage at
Oakland Coliseum,
December 16, 1978
Picture by Dennis Callahan

In the picture are Ronnie Montrose
and Bill Graham at KYA

The Cars at KYA Radio

Les Garland on the left,
Elliot Easton of The Cars
on the right.

Motown Convention in Las Vegas with Diana Ross

In 1979, I rented a Blue and Gold fleet ferry boat and took a load of radio listeners and promotion men out on the San Francisco Bay to promote the release of Jay Ferguson's *Shakedown Cruise* single.

Rear Left to right: Dino Barbis, Shelter Records, Dave Sholin, KFRC Music Director, (?), Lou Galiani, Warner Brothers, (?), Ron Lanham, Elektra Records, Les Garland, KFRC Program Director, Bill Thompson, Manager of Jefferson Starship, (?), Johnny Barbis, Sean Conrad, Chuck from WEA.

Front row: Mary, KFRC Promotions Director, (?), Fred Winston, Disc Jockey.

CHAPTER 19

"IT'S STILL ROCK AND ROLL TO ME"

In spite of that one night of out-of-town infidelity, and with Darla five hundred miles away, Birgitta and I patched up our marriage once again.

Although I was enjoying the record business, I couldn't see it in my long-term future. I was a radio guy. As 1979 was coming to a close, I heard that KCBS-FM was looking for a new PD to orchestrate a format change from disco to, well, something else. The FM division of CBS was a failure in just about every market, and the stodgy, conservative corporation didn't make it easy for someone to put together a winning format.

The company sent in one of their promising management stars, George Sosson, who had been the sales manager of one of their Philadelphia stations. As the new General Manager, the first thing George did was put out the call for a new PD.

My first meeting with him was at Mangia Mangia, a restaurant and watering hole on the ground floor of the Embarcadero One building in the financial district of downtown San Francisco. KCBS AM and FM were located on the thirty-second floor of Embarcadero One, which made Mangia Mangia a natural hangout for the CBS crowd.

George was short, about five foot five, with a desire to succeed in his eyes like no one I had ever met before. He was a teetotaler with only one vice, he liked the ladies. More than one female employee had been hit on by him at one time or another, including Darla while I was out of town.

I presented my case to him that day and was delighted to find I was hired at a hefty pay increase over what I was making in the record industry. I had been let go by ABC and was now an employee of its biggest competitor, CBS. In New York City, the CBS Building, also known as Black Rock, was across the street from ABC. I had been in the ABC Corporate Headquarters in New York while with KSFX in 1976, then three years later, got to walk across the street right into Black Rock.

George and I brainstormed a format we felt no other station had ever tried before. We came up with The Hits of Now & Then. Today it would be considered classic rock carefully mixed in with current music. We played songs in a two-to-one ratio of two hits to one oldie. We created and frequently aired one-hour specials devoted to the careers of former super groups like the Doors and Creedence Clearwater Revival. One weekend, we played nothing but Beatles records for seventy-two hours straight. No one had ever done that before, and the response was overwhelming. Regardless of where anyone went that weekend, if there was a radio playing, chances were it was tuned to CBS-FM.

The first order of business was to build the extensive music library we would need to match our new format. We had to physically locate all that vinyl. Prior to my arrival, when the disco format was thrown out at CBS-FM, the PD and his music director were out as well. I had to do the search alone.

Interestingly enough, just after I left KSFX and the disco format in 1978, they changed to rock. When they did that, CBS-FM went disco. Here it is two years later and I was being hired by CBS to change the format from disco to rock. Go figure.

When the time came to switch formats, I came up with a novel way to do it. Back in 1965, while auditioning records for air play, I stumbled across one by an unknown group called The Novas. The song called "The Crusher" was so bad that it was good. It was written by Reginald Lisowski, a former wrestler, and was a hopeless stiff. For the KCBS format change, I decided to play just that song over and over and over again for three days straight. This would, in effect, drive away the disco-listening audience and give us a clean slate for the new format. (Just go to YouTube, search "The Crusher" by The Novas to see and hear what I mean.)

We had tongues wagging in San Francisco when we pulled this off. It worked.

ℰↄℰↄ

"Sean, line two," came the voice over the intercom.

I picked up line two.

"Sean, it's me, Darla," she said, and my body froze.

"Hey, Darla. What's happening?"

"Well, I heard you got the PD job there. Are you going to be hiring a new music director?"

"Well I, uh, as a matter of fact, yes."

"Hire me. You know what a great team we were at KSFX. We made that station a winner, and we can do that again, Sean."

Oh shit. We did work well together. She, like me, was willing to put in the ungodly hours it took to make a radio station successful in the ratings.

"Don't do it," the rational voice of the angel dressed in a white flowing gown on my left shoulder screamed into my ear.

"But, you guys worked very well together," the voice of sexual desire sporting a blood red outfit bellowed in to my right ear. "Darla is just what the doctor ordered. She's a musicologist, and c'mon, you can keep your personal lives out of the equation."

Angel in white: "But you're married. You have two beautiful daughters and a lovely home in Marin County. Don't do it."

Devil in red: "Aw, go for it."

Angel in white: "You're weak. You know you can't resist her. You do this, and your marriage is over."

Guess who won this battle?

When I arrived at CBS-FM, my friend Tom Haule was the news director. Tom and I met when he was dating Joanie Hulein, General Manager Don Platt's assistant at KSFX. In the late seventies, Darla and Joanie shared an apartment. They had a huge falling out that I could never understand. One day, I took Tom and our chief engineer, Ozzie, to lunch at the Holding Company, a burger-and-fries joint a few blocks from the station.

"So Tom, I hired a music director," I said.

"Oh yeah, anybody I know?"

"As a matter of fact, you do. Darla."

Long pause. "You've got to be kidding me. You're nuts."

"Look Tom, I know something happened between you, her, and Joanie, but that's not my problem. Darla is a great music director, and she's been hired."

"You're making a big mistake," he said. "You'll be sorry."

In the back of my mind, I knew he was right. I knew I had ulterior motives in hiring her. The angel in the white flowing robes got it right. In no time, we were involved again.

Throughout all my years with Darla, most of my closest friends disliked her. She did know how to push people's buttons. Some of the support staff had difficulty working with her because they felt as if she talked down to them. I sometimes felt as if she talked down to me, too. Her superior attitude was a consistent part of our relationship.

Darla and I worked into the wee hours for months, building the music library. Many a night, I would stay at her apartment on Franklin Street and tell everyone else I was sleeping on my office sofa.

My life with Birgitta was on shaky ground, but she handled my infidelity with amazing calm. She was the steadying force in our family.

With Steve Perry

Darla Jenson and George Sosson

With Darla, J. Parker Antrim,
& members of Talking Heads
in the CBS-FM studios

With Darla and Patty LaBelle
at Sinbad's in the Embarcadero

CHAPTER 20

"IMAGINE"

L ate one Saturday night, as Darla and I continued to build the music library, my on-air jock at the time, Jim Bridger (Jim Hill), came in to my office.

"Hey, Jim. What's up?" I asked.

"You guys need a break. I got a joint. We could go smoke it out in the stairwell."

"Great idea," I said. "Let's do it."

We were on the thirty-second floor of the Embarcadero One, and there were no balconies or windows that opened. At two o'clock in the morning, lighting up in the fully enclosed fire escape stairwell, there was no chance of getting busted. We opened the door, walked down a few steps, sat down, and lit the joint. Jim had a long record on, so he had about five minutes to relax.

Bam! Someone fired a shotgun. At least that's what it sounded like as we lit up that doobie. The sound echoed off the barren cement walls, ceiling, and floors of the stairwell. We all jumped and then laughed at our nervousness. I was just beginning to feel the familiar high from the pot, when Jim coughed and sputtered.

"Son-of-a-bitch," he yelled. "The door."

Suddenly, I realized that the sound we heard was the door slamming shut behind us. We were locked in the stairwell.

"Shit. There's only a few minutes left on that record," he said. "How are we going to get back in?"

We started running. Down. Down. Down. Thirty-two flights of stairs. It was our only option. Get down those steps as fast as possible, hustle over to the elevator, and get back up there. We did, but for about ten minutes after the record ran out, it was as if we were off the air. That's how long it took to get back up there.

It wasn't the last time we smoked weed out there, but from then on, we always had something to prop the door open.

In the last several weeks of December 1980, Darla left for New York City on a two-week vacation. By this time, I was spending the night at her place more than my own. I sure didn't want her to go. Since we weren't officially a couple, I had no right to complain, and her desire to see old friends overpowered any concern she may have had about how I felt. While she was gone, I spent time with friends and just hoped she wasn't getting involved with questionable people again.

My co-worker, Les Isralow. was from New York City and fit all the stereotypes of the typical Jewish guy from the East Coast. In his shrill, high-pitched voice, he would say *I sore,* instead of *I saw*, *Let's play cads,* instead of *Let's play cards,* and *Paak the cah* instead of *Park the car*.

Before the Isralow boys migrated to San Francisco in the mid-seventies, Les's older brother Eric had taken solace in the world of music. His newspaper columns on the history of rock and roll, penned under the name of Dr. Rock, were gaining him some fame, but nothing like a road trip that Eric and his friend, Richard Boyle, took.

One day on a whim, loaded on beer and grass, the two of them decided they would drive their car to Central America. They drove right into the middle of a bloody revolution in San Salavador. Years later, in 1986, they were portrayed in the movie *Salvador* that Oliver Stone had made about their escapades. It was a true story, and Eric Isralow actually did do it. He was played by James Belushi, and Boyle, by James Woods.

Les, on the other hand, sold radio advertising and became known as the barracuda, a loving term assigned to him because of his take-no-prisoners approach to the sales game. Relentless in the pursuit of business.

I was over at Les's apartment on the night of December 8, 1980, with Dr. Rock and a few other people, watching TV, as the tenth season of Monday Night Football headed into the final weeks. We were partying it up as the Miami Dolphins battled the New England Patriots in a game that ended in Miami's favor, 16 to 13. But we never saw the end of that game. With three seconds left in the first half, Howard Cosell made a startling announcement.

"An unspeakable tragedy conferred to us by ABC News in New York City, John Lennon, outside of his apartment building on the west side of New York City, the most famous, perhaps, of all of the Beatles, shot twice in the back, rushed to Roosevelt Hospital…dead on arrival. Hard to go back to the game after that news flash, which in duty, we have to do. Frank," announced Cosell.

"Wait! Quiet you guys. Did you hear that?" I said to the revelers.

"Hear what?" Les shouted.

"Cosell just said John Lennon is dead. Oh my God!" I cried. "We need to get down to the station. I'll call Steve Garland and tell him to start playing nothing but John Lennon records."

"Sean, why don't you put my brotha on the air?" Les suggested. "I mean, c'mon, he's Dr. Rock. He'd have all kinds of stories to share."

"All right. Let's get going."

We all jumped in the car and drove to the station in a very quiet and somber mood. The reality of the situation was hitting us hard. John Lennon had just come out of retirement and was riding high on the success of his solo LP, *Starting Over*. It was gut wrenching to accept the fact that he was dead, and that a rumored Beatle reunion would never happen now.

When we got to the station, the phone lines were lit up.

"Sean, people are calling," Garland said. "They're crying. They're in shock. What should I do?"

"Let's put them on the air and let them express their sorrow," I told him. "I'm going to put Dr. Rock in the studio with you. The two of you talk to the listeners live on the air between songs."

I sat at my desk, listening in total shock when I noticed my private line was ringing, which was weird since it was so late in the evening.

I picked up the phone.

"Sean, its Darla. I'm in Central Park. You can't imagine what's going on here. Thousands of people are walking around in a daze, holding candles, and crying. It's cold and dark, but people are everywhere."

"Hold on," I said. "I'm going to put you on the air with Dr. Rock and Steve. Tell the listeners what you're seeing."

For the next few days, Darla made regular calls to the station from phone booths around Manhattan Island. As an eye witness, she reported what she was seeing as fans openly mourned the loss of a Beatle and held candlelight processions in his honor. Having Darla there was a godsend for the station. It was a one-in-a-million chance that someone on our staff would be in New York City at that historical moment.

That evening, and for the next few days, CBS-FM, along with hundreds of stations worldwide, devoted all their programming to the death of John Lennon. The following Sunday at noon, CBS-FM, along with most other stations across America went silent for one full minute in remembrance of John. At the end of that minute of silence, we all began playing "Imagine." It was a very moving and emotional experience. I can't listen to it today without remembering that moment.

Some of the Gang. In the picture: Steve Garland,
J. Parker Antrim, and Jim Hill.

CBS-FM billboard

NON-STOP BEATLES

Friday • Saturday • Sunday
June 26 • 27 • 28

Historic first on radio! 72 hours of non-stop Beatles...commercial free. Meet the Beatles through their music, interviews, live concert tapes, commentary, album cuts and TV sound tracks **without interruption**...not even for newscasts! Just solid Beatles midnight Thursday to midnight Sunday. "The Beatles Revolution – A 72 Hour Ticket to Ride," a radio first **commercial free** from KCBS • FM 97 with special thanks for their support to Serramonte Center, Thrift Town and The Camera Company.

KCBS·FM 97

We still play their songs.

Chronicle   6/25/81

Newspaper clipping about The Beattles

CHAPTER 21

"AND THE WALLS CAME TUMBLING DOWN"

My relationship with Birgitta was coming to an end. We both agreed the marriage was over. She started seeing someone else, and Darla and I became more and more entwined. We sold the house, and I got an apartment on Greenwich Street in the city.

My daughter Rhonda moved in to an apartment with her boyfriend, Mark. She had inherited my strong work ethic, and within a year, at age 16 became a manager of a year-round Christmas store, located on Pier 39 in San Francisco. Robyn moved in with Beth, who had re-married and lived in Marinwood.

I plunged even deeper into the disastrous relationship with Darla. We worked together all day, then were together all night. Darla was an expert at making me feel insecure, and I became jealous.

During this tumultuous time, we conducted an on-air contest in which the winner and a friend would get chauffeured in a limo to a Todd Rundgren concert at the Greek Amphitheatre in Berkeley. Rundgren was popular at the time with his hit, "Can We Still Be Friends," and listeners clamored to win the contest. The promotion was stitched together by Les Isralow in order to gain the lion's share of the advertising budget from Bill Graham Presents, who put on the event.

The day of the concert, Darla and I drove to Beth's apartment, just north of San Rafael, to visit Robyn. Darla and I were squabbling and started partying with my ex-wife and her new husband. Around 5 p.m., we drove back over the Golden Gate Bridge, got in the limo with the contest winners, and escorted them to the concert.

The young married couple who had won the contest were pleasant enough and excited about seeing Rundgren and doing a meet-and-greet after the show. We had the limo driver stop at a 7-Eleven, for snacks and champagne on the way, and fueled by additional alcohol, the hostility between Darla and me escalated. I'm sure the unsuspecting winners felt nervous and uncomfortable being so up-close and personal to this side-show.

Pulling into the back stage area of the Berkeley Greek Amphitheatre, our duties of escorting the winners around backstage were taken over by Bill Graham staff members. As everyone backstage enjoyed the loud music and the rock and roll ambience, the antics going on between Darla and me went mostly unnoticed. But not to Danny Shore, the manager of the event for Bill Graham. We were not physically, or emotionally, in any condition to be representing the radio station at this event. At one point, I pushed Darla out of the way in total frustration.

"Okay, you two," Danny yelled over an ear-piercing guitar solo, "I think it would be best if you both left this event now. Just have the limo take you home, and I'll have the driver return to take the winners home after the show."

At around midnight, we left the concert. Darla sat on one side of the limo, I sat on the other. She grabbed a wine glass full of stale champagne and drained it. The limo driver was driving us to our apartment via Broadway in North Beach.

As we crawled through the slow moving traffic, we drove past the Carol Doda Strip Club on the corner of Columbus and Broadway. The streets were filled with tourists and punk-rockers who were attending a show at the Mabuhay Gardens, a new age night club on Broadway. Just as we were slinking past the Mabuhay, Darla threw the wineglass toward the inside of the windshield, and I ducked as it exploded into millions of pieces. She opened the door, jumped out, and bolted into the Mabuhay.

I was shocked. So was the limo driver. We continued through bumper-to-bumper traffic towards the Broadway Tunnel when I jumped out, too. I ran toward the club to try and find Darla, because she had the key to our apartment. I just wanted to go home and sleep it off.

The Mabuhay, a Filipino night club, was Darla's second home. Owner, Ness Aquino, had turned the club over to Dirk Dirkson, the punk-rock version of Bill Graham. During its ten-year history, groups like the Sex Pistols, Devo, The Nuns, Romeo Void, The Ramones, Iggy Pop, The Dead Kennedys, and the Dammed played there. One local punk band called Nuclear Valdez made up of San Jose DJ Don Wright and his friend Eddie Hoyt, played there many times with their biggest fan, Darla in the crowd.

Everyone knew her at the Mabuhay. Those were her people. This was her world. I hated all that shit, but like a moth to a flame, I kept pursuing this woman who, other than radio, I had nothing in common with. I could not control the strong physical attraction I had for her.

Neither of us had money with us that night, but the bouncer knew her and let her in without paying the cover charge. I, on the other hand, was stopped at the door.

"Come on man," I said to the gorilla guard. "You're closing in less than an hour. All I want to do is get my key from her."

"Sorry," he barked. "No dice."

"Are you serious? I'll leave as soon as I get it," I promised.

"No. My job is to collect the cover charge."

This guy loved his job and the power that came with it, so there was no way I was going to get through that door without the *dinero.*

It was one o'clock in the morning, and the club would keep rockin' until two a.m. All I could do was loiter around the strip clubs and adult movie theatres and wait till the lights in the Fab Mab came up. As the club emptied out, I was permitted to enter. The bouncer was probably out back banging some green-haired punker-chick.

I found Darla leaning against the wall with her eyes closed. "Darla, we have to go home now," I pleaded. "Come on, let's go."

"Leave me alone," she slurred. "You pushed me. I'm not going anywhere with you."

"You've got the key, and we have no money. Come on, let's talk about it when we get home."

We eventually left the club, and I hailed a cab. When we got to the apartment, I ran upstairs to get money to pay him.

The next morning, lying in bed trying to clear the cob-webs, we both started getting flashes of what had transpired the night before. Fear and panic flooded in.

"Sean, what did we do last night? Oh my god."

"I think we fucked up big time," I replied.

All the bad feelings we had experienced the night before were gone. Now it was us against them. We continued to take inventory only guessing what we would face Monday morning at work. Then, the phone rang.

"Sean, this is Les. How'd it go last night?" he inquired cheerfully, expecting to hear how great the evening had gone.

"Uh, well, not so great, Les."

"What do you mean?" he squeaked. "What happened?"

"Les, I've got to go. We'll talk about it tomorrow."

The whole thing blew over, and we went back to business as usual. The times were so different in the late seventies. Obviously, some people were given way more slack to screw up than they should have been. We found out a few days later that the woman who won the contest was the daughter of Diane Feinstein, who at the time was the Mayor of San Francisco.

လ၁လ၁

The format we developed at CBS-FM was a hybrid made up of oldies, current hits and album cuts. Within the first two rating periods, we managed to increase listeners from 70,000 to 180,000. And that was directly connected to increased advertising dollars for CBS.

As a perk to the increase in ratings, Darla and I frequently attended rock concerts at the Old Waldorf, Fox Warfield, Fillmore West, the Great American Music Hall, the Cow Palace, and the Oakland Coliseum. We went backstage at most all of them and met many rock stars.

I put Darla on the air so she could perform two jobs and save the station a salary. She became an on-air music director to help cut down on overhead. She had a warm, friendly voice and sounded great.

My air staff at CBS-FM included J. Parker Antrim, Jeff Dean, John Mack Flanagan, Steve Garland, and weekenders, Dirk Robinson, Peter B. Collins, and Jim Hill.

CBS-FM continued to grow in the ratings, but we never reached the audience levels the company was expecting. By late 1981, CBS opted to change the format, its fourth change in the last five years. I was replaced by Dave Roberts, who was more of a statistician than a program director. I was back on the streets again. A few days later, Darla was also fired.

My style of gut-feeling programming was out, and Dave Roberts' research was in. Each and every song was first played for a focus group, and any song that didn't appeal to the masses was nixed. Dave did turn the station around with a call letter change to KRQR (the Rocker), and a format change to balls-against-the-wall hard rock.

I was ready to toss in the towel anyway. It was time for a drastic change in my life. I was frustrated being in the major market program director's hot seat, when I, as the PD, had only so much control over the success of the radio station. We were handcuffed by a management, who insisted on meddling in programming, and hamstrung by a promotion budget that was too small to compete.

After work at Mangia, Mangia
in Embarcadero One

At a party with Bruce Hix
from Warner Brothers Records

Publicity Still

Mabuhay Gardens poster

CHAPTER 22

"(JUST LIKE) STARTING OVER"

In late 1981, after Darla and I were axed at CBS-FM, we moved into an apartment in Pacific Heights. We figured we could share the rent and survive until the next gig came around. With our severance pay, we flew to Hawaii for a vacation on Maui. We were far from being happy together, and our five-year long on-again, off-again co-dependent, tempestuous relationship crashed to a final end.

It was then that I made the decision to get out from behind the microphone and into the sales game. After all, every radio station salesman I ever knew lived the life of Herb Tarlick on WKRP in Cincinnati. Get to work at 10. Go to lunch at noon. Drink your lunch. Make a few sales calls. Return to the station. Leave at 4. Besides, they didn't have to worry about ratings, and they made more money than the jocks. It was a no-brainer.

My younger brother, Jim, was sales manager of KBLQ radio in Logan, a town of 25,000 in northern Utah. When he offered me a job selling advertising, I jumped on it. He knew what was going on in my life and that I needed a fresh start.

"Ron, why don't you come over here and sell for me? I'll put you in the trenches, and you can learn sales from the ground up."

"What kind of money could I make starting from scratch on the bottom like that? I'm broke and need to pay my bills, too, you know."

"I talked to the GM and he said, to make a living, you'd also have to do morning drive. What do you think?"

"So I get up at 5 a.m., go on the air at 6 a.m., get off at 10 a.m. and go sell advertising?"

"Uh-huh."

"Well, I guess I could do that. What the hell? But I doubt I'd stay long. As soon as I thought I was ready, I'd be looking for a job back here in California closer to my girls."

I had run out of chances, and my job in Utah would be the start of a new life for me. No more crazy hours. No more driving under the influence. No more drugs. I hooked up my un-restored 1965 Ford Mustang that had transmission problems and a heater that didn't work, to the back of a U-Haul. I was going from king of the hill to low man on the totem pole, *again*. It was a humbling experience. Logan became the sleepy little country town where I shared an apartment with my brother and focused on work.

I stumbled and stuttered my way through the process of learning sales from ground level, and I was motivated to make money. Driving around in sub-zero weather in a car with no heater will do that to you. It'll build a fire under you, so to speak. During that winter of 1982, the temperatures in Logan never went above zero for one full month, not even in the middle of the day.

I took a night class in basic accounting so that I could learn how to budget and rebuild my credit. As I sat in that class, I realized that I had been handed a final opportunity. This time, I was not going to blow it.

At the end of 1982, I knew it was time to get back to California. My friend Dirk Robinson, who had worked for me as a jock at several radio stations, including morning man at KYNO, had moved to Santa Cruz on the northern tip of the Monterey Bay. He was living there and working for a daytime AM station called KMFO, which was in the Santa Cruz suburb of Aptos. I loved Santa Cruz and was familiar with it because, in the early '70s to escape the Fresno heat, Beth and I would take the girls to the Santa Cruz Beach Boardwalk.

"Hey Sean, KMFO is looking for a GM. Why don't you go for it?" Dirk suggested. "You could be my boss again."

I applied for the job, and starting in January of 1983, I was back in California and living in paradise. KMFO was owned by the Wrathall brothers, two guys without the slightest clue how to run a radio station. They had inherited it from their father, who was a brilliant radio engineer in the '40s & '50s.

The boys gratefully turned over the station to me to try to make it profitable, but with a zero budget to operate with. Hell, I didn't care. I just wanted to get back to California. They paid me $300 a week and all the trade I could generate. We ran syndicated news-talk programs, supplied by CNN, and talk shows that originated at ABC as our format, with a few local newscasts to glue it all together.

The first thing I did when I got there was to hire a sales staff. For sales manager I brought in the barracuda Les Isralow who was unemployed and still living in San Francisco. As account executives, we convinced my buddy Dirk Robinson to give up the microphone and put on a sales hat. Then we hired a young kid fresh out of Chico State College, Chris Chidlaw, who was full of piss and vinegar and anxious to start a career.

After two years of banging my head against the wall trying to turn that turkey around, I threw in the towel. It became obvious that an AM daytime station at 1540 on the dial didn't stand a chance of generating a profit. It would be insanity to stay there, and I knew it. But I did continue to master the art of creative selling. You *had* to be creative to sell a radio station with no listeners.

In October of 1985, I was hired at legendary KDON AM and FM Radio in Salinas as a bottom of the pecking order new sales guy. They gave me the Gilroy and Morgan Hill territory, and every day I'd get in my 1976 Chevy Impala with the leaky T-top and make the drive from Santa Cruz over highway 152 to Gilroy to knock on doors and sell, sell, sell.

If you've ever driven that hill, you know what a treacherous drive it can be, especially in rainy weather. Imagine navigating those hairpin turns in the rain with water dripping on you from the T-top as you hold an empty coffee cup in one hand, trying to catch the drops before they made stains on your dress slacks, while, with your free hand, keeping the car on the road. It's not easy, but I was hungry and started bringing in advertising dollars from the highly neglected territory by the boat loads. Within six months, I was promoted to sales manager of the entire operation.

My three years of selling advertising in the trenches had paid off. With my knowledge of programming and now, sales too, I started breaking all sales records month after month and began to make really good money. I re-built my credit, bought a new car, and dug myself out of the hole. I would never again self-destruct as I had done numerous times before in the first forty years of my life. I finally *got it*.

In 1985, while reaping the benefits of a successful sales effort, I decided to start a side business and to develop a second source of income. I bought two old, beat-up Ampex tape recorders from KMBY-FM in Monterey, a radio shack microphone and mixer, and set up a little studio in my apartment. I called my new business The Spot Farm, and I was off to the races. I began to lure local advertising agencies to my studio for their production needs. In 1986, I did my first session with the Santa Cruz Beach Boardwalk for their "Bands at the Beach" series and other annual events like the Clam Chowder cook off. Had I not taken the chance to start a second business, my life would be nowhere as comfortable as it is today. I thank my lucky stars for that decision.

Between being a sales manager of a successful radio station and my new found second stream of income, I was able to buy a house in Aptos. This would be the fourth home I would own, but this time, my name was the only one on the title. Having been a summer cabin in the '60s, it had two bedrooms upstairs, a one-car garage, and was situated on the bank of a small stream with giant, two-hundred-year-old redwood trees all around. As a matter of fact, one of these majestic monoliths protruded right through the back deck where I could stand on a beautiful Sunday morning, drinking a cup of coffee, and literally be a tree hugger while listening to the trickling creek below. Eventually, I added on a third bedroom and another bath and a half and sheet rocked the garage which became the home for Spot Farm Studios.

Around the middle of 1990, I saw my first locally inserted TV commercial on United Artist Cable which served the Santa Cruz area. It blew me away. I immediately saw my future and applied for a job as a salesman. Again, more divine intervention. I visualized how local cable ad sales could put a dent in radio advertising budgets, and I was right.

On January 2, 1991, I became one of three sales people for a cable system that had just started selling locally inserted advertising only six months before. All of Santa Cruz County would be my oyster.

Once again, Russ Terrana and I crossed paths when he came to work for us as a TV commercial producer. Russ would make the commercials, and I would sell them. We did just that.

In 1992, I got a call from Les Isralow. He was now the General Manager of KDON, and I thought maybe he was looking for a new PD and Music Director.

"Sean, did you hear about Darla?"

"No, did you hire her?" I asked.

"She died," he said.

"What? How could that be? She's only thirty-nine."

"Someone said it could have had something to do with drugs," he told me. "It happened a few days ago."

I had talked to Darla only a few weeks before when she called from Southern California. She told me she was clean and sober and going through her steps. As a volunteer to help promote a public service campaign, she wanted to know if I'd do some free production for her. I said yes, but never heard back from her. I figured the whole thing must have fallen through.

As I look back on it, there, but for the grace of God, went I. The party lifestyle we both shared to the max may have had something to do with her death, although I never found out for sure. With both of us now more mature, I had hoped the day would come when we could apply some closure to those insane years. It didn't happen.

With the barracuda, Les Isralow, and talk-show host,
Mike Jackson, while at KMFO

CHAPTER 23

"THE LONG AND WINDING ROAD"

By 1996 I was a top salesman, winning national recognition and trips to Cancun and the luxury five-star Broadmore Hotel in Colorado Springs to accept my awards. I was making a six-figure income and enjoyed every minute of it. At the same time, my recording studio was blossoming, too.

My daughters are both grown with families. Rhonda is manager for five Clear Channel radio stations in the Monterey Bay area, including my old station, KDON. Robyn recently moved to Alaska to start a new chapter in her life. I'm still in touch with Beth and Birgitta, and my daughters are close to both of them.

In 1994, I met Lisa, my soul mate, at TCI Cable. We married two years later, and she remains a peaceful, grounding force in my life.

In 2008, we purchased a 40-foot top-of-the-line, class A Country Coach motorhome. We now work, write, travel, and play in it full time. The Spot Farm has become an advertising agency, and we produce radio and television commercials while on the road.

I am thankful I worked in radio at the time I did. Today, companies own hundreds of stations all across America. With technology, they can run a radio station with a third of the staff it took to run one back then.

Smaller staff means fewer new announcers entering the workplace at an early age, which decreases the available talent pool. Jocks have to do the workload of two or three people, work longer hours, generally for less money. Being a jock today is more like working in a factory. It's just a job.

My jock staffs in the 1970s were always tightly-knit groups. We were a brotherhood all focused on the same goal: increasing the size of the listening audience. Sure, there were rivalries and ego clashes, but no one could pick up on it over the airwaves. Being promoted to a better air shift, which usually meant a bigger paycheck, was especially hard on the jocks who were passed by. But no one had anything to complain about when it came to working conditions. In those days, after an air shift, a jock might spend an hour in production and then be out the door. A 30-hour work week with full pay and benefits was not uncommon.

The absolutely most enjoyable time I spent as a program director was the three years I worked for Bill Drake and Gene Chenault in Fresno. They put little pressure on me, and I knew I had their absolute total support. They treated me with great respect and made me want to jump out of bed in the morning and go to work for them. I owe those guys my professional life.

What I learned from these two radio legends in Fresno launched me on a journey that took me to Chicago, Los Angeles, Honolulu, and San Francisco.

On November 29, 2008, Bill Drake died of cancer at the age of seventy-one. I flew to LA and attended the services, which were held at the Little Brown Church on Coldwater Canyon in Studio City, California.

Drake-Chenault workers who hadn't seen each other in years re-united once again to honor the life of a radio legend. Speakers included Charlie Van Dyke, now an ordained minister, who conducted the services. Those who stepped up to the podium to tell Drake stories included Bill Watson, Bernie Torres, Les Garland, Ken Levine, Charlie Tuna, and Gary Owens.

Gene "Pappy" Chenault did not attend the services because of ill health, but I could feel his presence. Without him, none of us would have been gathered in that room sharing a lifetime of memories.

When I heard two years later that Gene had passed away at the age of ninety, I thought back on how my personal life was such a mess while I was his PD. I made some really bad choices that today would get one fired. Gene always stuck by me and was my biggest fan. I busted my ass for him, and he rewarded me not only financially, but with the emotional support I so desperately needed in my troubled twenties. He treated me like his son. The world of radio is a better place because of Gene Chenault.

I am now going to go have a Tanqueray with a twist, his favorite adult beverage, and I am going to count my blessings for having known him.

The music of those times will continue as long as there are people to hear it. It will connect us to the memories as it always has. In the late '60s to early '70s there was a time when less than 10 program directors at any given moment were employed by Drake-Chenault. I am honored to have been one of them.

Little brown church in Studio
City where Bill Drake's
memorial services were held

With Charlie Van Dyke
at the memorial services

Bill Drake

Gene Chenault and wife
Suzy in the '70s

Author's Note

I'm not saying that everybody in radio did what I did back then...but this is just how I handled life. Some people may also have different recollections of certain events portrayed in this book. There are two sides to every story, after all. Also, some of the names have been changed to protect the privacy of the individuals involved.

Acknowledgements

I especially want to thank the many crazy, wacky, eccentric, goofy, slightly off-center, funny, bizarre, boisterous, mischievous, fun-loving, hard-working, neurotic, obsessive, compulsive, outgoing DJs I worked with through the years. Without a strong desire to join your club, I may have become a house painter, or an appliance dealer, or any of a million vocations that would have never allowed me to have so much fun in my life.

I want to thank my dad, Clyde Copeland, for believing in me from day one. Because of your abundant love, encouragement, and support, I got to experience an exciting, and anything but boring, career in major market radio. Even more importantly, the way you applied the golden rule to your dealings with your fellow man throughout your life was a lesson I have tried to follow throughout mine. "Son, be the best, not one of the best," still rings in my ears.

My mentor, Bob Holliday, for his direction in my early days in radio.

Thanks, also, to Don Branker, Bill Drake, Gene Chenault, Steve Randall, Roger Turnbeaugh, Jim Davis (Big Jim Edwards), Allen Shaw, Russ Terrana, Beth Widdowson, Riley Cardwell, Glen Martin, and Jeff Kauffer.

A huge thank you goes to Bonnie Hearn Hill, for her generous help and inspiration, as well as The Fridays Writing Group (Bonnie, Hazel Dixon-Cooper, Christopher Allan Poe, and Dennis C. Lewis) that I was privileged to sit in on.

RADIO STATIONS

KCBS FM San Francisco: The KCBS FM story dates back to 1978 when a four way station switch moved the station frequency to 97.3 FM. In October of 1978, a new format was tried, "The Hits of Now And Then," that played oldies from 1964 to the present day. In January 1979, the station tried a disco format. It didn't work. Many formats have been tried since. On July 12, 1991 in Los Angeles, KODJ adopted the KCBS-FM call letters used in San Francisco. On September 1, 1993, KCBS-FM Los Angles dumped the oldies music format in favor of classic hits music, focusing primarily on the 1970s. On March 17, 2005, the classic hits music format was dropped in favor of its present music format of adult hits branded as Jack-FM or The JACK – Los Angeles.

93 KHJ Boss Angeles: In 1964 two California radio programming pioneers, Bill Drake and Gene Chenault, modified the Top 40 formula and called it "Boss Radio." Drake and Chenault introduced their new format in April 1965 on KHJ. Within a few months, the "Boss Radio" format had brought KHJ to the top of the Los Angeles market. On November 7, 1980, KHJ switched over to a country music format, which lasted barely two and a half years. In April 1983 they switched back to a top-40 style format, declaring on the air that "The Boss Is Back." On January 31, 1986, KHJ abandoned its classic call letters and adopted those of its FM outlet KRTH. They began playing "Smokin' Oldies" from the first ten years of rock 'n roll. This format lasted until the late 1980s. In the early '90s they went Spanish, changing their call sign to KKHJ with an all-Spanish news format, "La Ranchera." On Wednesday February 27, 2013, the historic KHJ towers, at 5901 Venice Blvd., came tumbling down, after four days of planning to make sure they came to their final rest safely in the residential neighborhood of Fairfax. There were many of us old DJs who suddenly felt a whole lot older.

KIQQ (K-100) Los Angles: KIQQ debuted in 1972 as a lighter top-40 format. In 1973, Bill Drake and Gene Chenault, just shown the door at RKO, cut a five-year deal to program KIQQ, but Drake and his top talent (Morgan and Steele, who both quit KHJ within weeks of Drake's departure) were bound by non-compete clauses in the KHJ contracts, keeping them off any other Los Angeles stations for six months. So lame duck KIQQ (which used only the call letters) limped along until the fall of 1973, when Drake, Morgan, Steele, Billy Pearl, Jerry Butler, and Humble Harve kicked off K-100. In the early '80s, the format was changed to Transtar's "Format 41" satellite AC service, keeping the call letters KIQQ but using them only for the legal ID once an hour. After several ownership changes, today 100.3FM is KSWD 100.3 The Sound. The station currently broadcasts a wide-ranging *classic rock* format. *Note: Drake-Chenault was sold and eventually dissolved in the mid-1980s, but their radio specials are still available from a variety of sources.*

WKNR Detroit: WKNR officially launched on October 31, 1963, with the "Battle of the Giants," an attention-grabbing promotion that invited listeners to call in to vote for their favorite oldies. The station quickly gained momentum, and lasted until the spring of 1967. WKNR's dominance was challenged when CKLW-AM got a makeover, courtesy of Bill Drake and Paul Drew in April 1967. The Big 8 became the number one Top-40 station in the region. On April 25, 1972, "Keener 13" signed off to the sounds of "Turn! Turn! Turn!" by The Byrds, and changed to an *easy listening* format as WNIC, simulcast with its FM sister station AM 1310 simulcast for a short time until the decision was made in 1977 to revive the "Keener 13" brand name on its original frequency with an adult-oriented Top 40/Oldies mix and a new call sign, WWKR. The second version of "Keener 13" did not have the success of the original, and by 1980, AM 1310 was back to simulcasting. Since late 1986, AM 1310 has tried several other different formats, none of which have attained lasting success, and has been in and out of simulcasting.

KTKT Tucson: In December 1949, Tucson heard its fifth station, KTKT (at AM 1490) and it became Tucson's only rock and roll radio station. By 1956, the station frequency was moved to 990 AM where it was nicknamed Color Channel 99, as color TV was the rage in the early 1960s. KTKT was programming the new rock and roll-Color Radio-Top-40 style music and news format which quickly moved daytime-only KTKT into Tucson's number one spot where it remained into the early 1980s. The Color Channel 99 bumper sticker was notorious, *Don't honk, listening to KTKT Color Channel 99.* From the '80s on, KTKT became a Spanish speaking sports radio.

WING Dayton: It was in the 1940s during the Big Band era and later with the advent of R&B-fusioned rock n' roll in the 1950s, when WING became Dayton's original hit music station. In the '60's Then High Flying WING was the theme of a high energy upbeat format with a jingle package to match, produced by PAMS Productions in Dallas, but it was in the mid-1970s that FM rock stations started to chip away at AM radio's Top-40 audiences. During this transitional time, WING began to soften its format to *adult contemporary* as "Adult Radio 1410." But in the 1980s and '90s, WING showed signs of listener burnout as even more listeners switched to FM. After a stint as a CNN Radio affiliate in the 1990s, and various network talk programs, it found its new niche in sports/talk. On May 17, 2007 Philadelphia-based Main Line Broadcasting announced the acquisition of Radio One's stations in the Dayton and Louisville market areas which included WING on September 14, 2007.

KPOI 1380 Honolulu Hawaii: KPOI was a legendary Top-40 station. It was the home of rock and roll and the *Poi Boys*, (Ron Jacobs, Tom Moffat and Dave Donnelly.) It was the #1 rated top-40 music station in Hawaii for most of the '50s, '60s, and the early '70s. Sadly, today there is no longer a 1380 AM frequency in Honolulu, but the KPOI legend lives on as KPOI FM 105.9, "The Rock You Live On."

KYNO 1300 Fresno: From 1957 and throughout the '60s and '70s, was a Top-40 station in Fresno, California under the ownership of Gene Chenault. KYNO was the testing ground for the "Boss Radio" format that would be adopted at such stations as KHJ, Los Angeles, KFRC, San Francisco, and KGB, San Diego. In 2008 KYNO was purchased by John Ostlund, owner of FM station KJWL. In April 2010 the KYNO frequency was changed from 1300 AM to 940 AM with a politically conservative *news/talk* format. In November 2012, KYNO moved to 1430 AM and changed to a '60s music format. That move takes KYNO back to its AM music radio days.

KSFX San Francisco: In 1974, the station veered towards a Dance/Soul-flavored format. During the late 1970s, the station had a brief run with a *disco music* format, then after several different calls letter and format changes, the station found its home at 103.7 FM, The Greatest Hits of the '60s '70s & '80s.

BUSINESS CARDS

Ron Copeland
Program Director

WDAI Radio
360 North Michigan Avenue
Chicago, Illinois 60601
Telephone 312 782-6811

93 KHJ

KMFO NEWSTALK 1540

SEAN CONRAD
GENERAL MANAGER

7600 OLD DOMINION CT. SUITE C APTOS, CA 95003   (408) 688-5904
                                                        (408) 475-5527

Sean Conrad
Program Director

KYNO

RADIO KYNO
2125 N. BARTON - FRESNO, CALIF. - 255-8383

CBS FM 97

Sean Conrad
Program Director

KCBS/FM 97
One Embarcadero Center
San Francisco, CA 94111
(415) 765-4097

M

SEAN CONRAD
West Coast Regional
Pop Promotion Manager

MOTOWN RECORD
CORPORATION

70 Dorman Street
San Francisco, California 94124
(415) 286-0959
(415) 924-2957 - Home

SEAN CONRAD

**ELEKTRA/ASYLUM/NONESUCH
RECORDS**
680 Beach St., Suite 452
San Francisco, CA 94109
(415) 441-6111

A Division of Warner Communications Inc.

K-POI
AM 1380

SEAN CONRAD
Program Director

COMMUNICO 1701 ALA WAI BOULEVARD, HONOLULU HAWAII 96815   808 941-0644

Sean Conrad
Program Director

KCBS-FM
CBS Radio, A Division of CBS Inc.
One Embarcadero Center
San Francisco, California 94111

(415) 765-4036

Sean Conrad
Program Director

KSFX Radio
1177 Polk Street
San Francisco, California 94109
Telephone 415 928-0104

SEAN CONRAD, CRMC
Certified Radio Marketing Consultant
General Sales Manager

Monterey Savings Tower, P.O. Box 81460, Salinas, CA 93912
Salinas 422-5363 • Monterey 649-1732 • Santa Cruz 688-9311

100.7 FM

COUNTRY FAVORITES
Sean Conrad
Santa Cruz County Sales Manager

(408) 662-3112
(408) 688-4401   P.O. Box 81380
Fax (408) 685-0847   Salinas, CA 93912

United Artists Cable Advertising
of Santa Cruz County
106 Whispering Pines Drive
Scotts Valley, CA 95066
(408) 439-5099 ext. 3410
FAX (408) 439-5064

UNITED ARTISTS

Sean Conrad
Account Executive

Aptos
Spot
Farm

Sean Conrad
(408) 688-4401

996 TROUT GULCH ROAD • APTOS • CA • 95003
A Santa Cruz County based Radio and Television Commercial Production Company

4647 S. Equestrian Drive
Sierra Vista, AZ 85650
(520) 803-7219
Cell (520) 227-8038
Fax 1-866-457-4128
www.spotfarmstudios.com
ron@spotfarmstudios.com

The **Spot Farm.**
Media Services

Ron Copeland

THOSE THAT LEFT TOO SOON

Dino Barbis: 2000

Gene "By Golly" Barry: 2001

Hal Davis: 1998

Mike Deeb

Dave Evans: 1997

"Skinny" Bobby Harper: 2003

Eric Isralow (Dr. Rock): 2011

Darla Jensen: 1992

Hal Lewis "J. Akuhead Pupule": 1983

Art Roberts: 2002

Ron Sherwood: 2005

Bernie Torres: 2011

Kris Van Kamp: Died in the mid-'90s

Wolfman Jack: 1995

John Yount/Big John Carter/Maxx Mahi Mahi: 2005

WHERE ARE THEY NOW?

Joe Angel: Radio play-by-play announcer for the Baltimore Orioles.

Jon Badeaux: Traffic Systems Manager for Entercom in Seattle, Washington.

Joe Bailey: Retired in Phoenix, Arizona.

*Johnny Bar*bis: Chairman of North America Rocket Music Entertainment Group. Most recently, he has embarked on a new role as manager to his old friend Elton John.

Dick Biondi: On the air at Chicago's 94.7 FM, True Oldies, weeknights 8-midnight.

Don Branker: Retired in Fresno, California.

Jim (Hill) Bridger: Finally escaped the KCBS-FM 32nd floor stairwell, and went on to become one of the first anchors at E Entertainment Television. Jim is eternally grateful to Sean for re-hiring him two weeks later under a different name, saving his career after Sean had to fire him at the behest of lawyers from CBS Corporate and the Mayor of Oakland.

Paul Cannon: Lives in Peoria, Illinois.

Dex Card: Formerly owned a four station chain in the Great Lakes area. Currently retired.

Riley Cardwell: Is now referred to as Ehu Kekahu. Cardwell—I mean Ehu Kekahu—is involved in the movement for a free and independent Hawaii.

Scott Cohoe: Involved with retail in Omaha, Nebraska.

Fred Constant: Owner of Diamond Mountain Vineyard in Calistoga, California.

"Big" Jim Davis: Vice President, Market Manager of Treasure and Space Coast Radio in Vero Beach, Florida.

Bob Dearborn: Retired in 2010 and lives in southwestern Ontario.

Les Garland: Les is the co-creator of MTV and VH1. He is still active in the music business, lives in Miami, Florida, and plays a lot of golf.

Steve Garland: Senior Marketing Consultant, Broadcaster, and CEO of The Jingle Lab at Entercom Sacramento, SteveGarlandMedia.com, TheJingleLab.com.

Dan Gates: Finally hung up his vocal cords, and is retired in Ohio.

"Uncle" Russ Gibb: Co-founder of the music discovery website GrokMusic.com.

Gary Granger: Founder of GatorsScore.com. He lives in Gainesville, Florida.

Birgitta Hallgren: Became successful in property management in Marin County.

Tom Haule: Anchor at CBS radio station KNX in Los Angeles, California.

Ed Hoyt: Realtor in Pebble Beach, California

Paul Iams: Founder of Ac·Rock, sings a cappella rock 'n' roll, performing everyone's favorites from the 50s through the 80s.

Les Isralow: producing radio shows on real estate investing.

Conrad M. Jimenez: Owner/Lead Designer, ART Conrad DESIGNS Fresno, California, Architectural Design Firm, responsible for the development of interior/exterior building finish schedules, to include contract, commercial and residential projects.

Ted Jordan: Involved in film and music production, living in Phoenix, Arizona.

Frank Kalil: Continues as a successful radio/television station broker operating out of Tucson, Arizona.

Jeff Kauffer: No longer "on the air"…is now "in the air." Jeff is now an airline pilot living in Seattle, Washington.

Barry Kaye: PD/afternoon drive at Country KVST in Houston, Texas.

Jim Kerr: Still rocking at Q104.3, New York's Classic Rock.

George Klein: Heard on Sirius XM Elvis Radio.

Ken Levine: His blog, kenlevine.blogspot.com, was recently named one of the BEST 25 BLOGS OF 2011 by TIME magazine. He is an Emmy-winning writer/director/producer/major league baseball announcer.

Captain John Lodge: Working at WTXU in Philadelphia.

John Mac Flanagan: Retired, living in San Francisco, California.

Glen (McCartney) Martin: Left radio in the 90s, and is now a financial advisor in the Seattle, Washington area.

Rhonda (Copeland) McCormack: Station Manager for Clear Channel Radio (KDON, KTOM, KPRC, KION) in the Monterey Bay area.

Harvey Mednick: Creative Director for Lemon Grove Press.

Joel Newman: President/Chief Creative Officer at www.shmek.net Shmek-U "The University Network."

Mike Novak: CEO and President of EMF, K-LOVE, and Air 1 Broadcast Networks.

Gary Owens: Currently the primary voice of the over-the-air digital network Antenna TV.

Lee Poole "Johnny Rabbit": Retired.

Steve Randall: Until recently, he did voice-overs and imaging full time. He is now retired, living in Fresno, California.

Bobby Rich: Program Director/Morning Show Host/Director Community Partnerships at Journal Broadcast Group, Tucson, Arizona.

Ron "Ringo" Riley: Retired after more than twenty years in television.

Dirk Robinson (Raaphorst): Lives with his wife and cat on their boat in Moss Landing, California. He still does occasional radio and television voice-overs.

Ron Samuels: Living in Houston, Texas, he is the owner of The Samuels Company.

Gary Sandy: Last known, he was doing musical theater.

*Alan Sh*aw: President and CEO at Centennial Broadcasting.

Bruce Shindler: Vice President/National Promotion at Universal Music Group in the Greater Nashville Area.

Dave Sholin: Has an afternoon show on country station KSJJ 102.9 in Bend, Oregon.

George Sosson: President of Radio Station Management, Inc., and is also associated with Rep Radio Corp. in New York.

Bill J. Stevens: A renowned actor, who has appeared in sixteen films, such as *All God's Children*, and *A Killing Affair*.

Jerry Stowe: Retired, living in North Carolina.

Ralph Terrana: Retired and living in Monterey, California.

Russ Terrana: Happily living in Santa Cruz, California with his wife. He still dabbles in music production.

Jeff Trager: Vice President of Second Octave Talent, a full service booking and tour management company in Petaluma, California.

Charlie Tuna: Has a weekend show on KRTH in Los Angeles, California.

Charlie Van Dyke: An ordained minister, and a major voice-over talent for radio and television. He lives in Paradise Valley, Arizona.

*Dave Van D*yke: General Manager Affiliate Relations at Radiate Media.

Bill Wade: Teaching at Lampson Business College in Mesa, Arizona.

Bill Watson: Retired, living in North San Diego County, California.

Clark Weber "Mother Weber's Oldest Son": Currently promotes *A Senior Moment with Clark Weber*, a pre-recorded program series for radio stations targeting senior health issues.

Beth (Castle-Copeland) Widdowson: Beth and her husband Rick own Widdowson Construction in Turlock, California.

Johnny Williams: Lives in Hawaii and hosts the premier radio website, 440int.com.

Steve Wrath: Account Executive with Univision in Fresno, California.

About the Author

Sean Conrad (Ron Copeland) began a career in radio at the age of thirteen. He spent more than thirty years as a disc jockey and program director for twenty-two radio stations in cities such as Chicago, Detroit, Los Angeles and San Francisco. He and his wife Lisa now create radio and TV commercials as they write, work, travel, and play their way across America in their 40-foot motor home.

www.ingramcontent.com/pod-product-compliance
Lightning Source LLC
Chambersburg PA
CBHW060236050426
42448CB00009B/1461